International Conference on Geological, Geographical, Aerospace and Earth Sciences

(AeroEarth 2014)

IOP Conference Series: Earth and Environmental Science Volume 23

Bali, Indonesia
11 – 12 October 2014

Editors:

F. Gaol
B. Soewito

ISBN: 978-1-63439-964-7
ISSN: 1755-1307

Printed from e-media with permission by:

Curran Associates, Inc.
57 Morehouse Lane
Red Hook, NY 12571

Some format issues inherent in the e-media version may also appear in this print version.

Printed by Curran Associates, Inc. (2015)

For permission requests, please contact the Institute of Physics
at the address below.

Institute of Physics
Dirac House, Temple Back
Bristol BS1 6BE UK

Phone: 44 1 17 929 7481
Fax: 44 1 17 920 0979

techtracking@iop.org

Additional copies of this publication are available from:

Curran Associates, Inc.
57 Morehouse Lane
Red Hook, NY 12571 USA
Phone: 845-758-0400
Fax: 845-758-2634
Email: curran@proceedings.com
Web: www.proceedings.com

TABLE OF CONTENTS

Preface

The 2nd International Conference on Geological, Geographical, Aerospace and Earth Sciences 2014 (AeroEarth 2014), was held at Discovery Kartika Plaza Hotel, Kuta, Bali, Indonesia during 11 - 12 October 2014.

The AeroEarth 2014 conference aims to bring together researchers and engineers from around the world. Through research and development, earth scientists have the power to preserve the planet's different resource domains by providing expert opinion and information about the forces which make life possible on Earth.

Earth provides resources and the exact conditions to make life possible. However, with the advent of technology and industrialization, the Earth's resources are being pushed to the brink of depletion. Non-sustainable industrial practices are not only endangering the supply of the Earth's natural resources, but are also putting burden on life itself by bringing about pollution and climate change.

A major role of earth science scholars is to examine the delicate balance between the Earth's resources and the growing demands of industrialization. Through research and development, earth scientists have the power to preserve the planet's different resource domains by providing expert opinion and information about the forces which make life possible on Earth.

We would like to express our sincere gratitude to all in the Technical Program Committee who have reviewed the papers and developed a very interesting Conference Program as well as the invited and plenary speakers.

This year, we received 98 papers and after rigorous review, 17 papers were accepted.

The participants come from eight countries. There are four Parallel Sessions and two invited Speakers.

It is an honour to present this volume of *IOP Conference Series: Earth and Environmental Science (EES)* and we deeply thank the authors for their enthusiastic and high-grade contributions. Finally, we would like to thank the conference chairmen, the members of the steering committee, the organizing committee, the organizing secretariat and the financial support from the conference sponsors that allowed the success of AeroEarth 2014.

The Editors of the AeroEarth 2014 Proceedings

Dr. Ford Lumban Gaol
Dr. Benfano Soewito

IOP Conf. Series: Earth and Environmental Science **23** (2015) 011001 doi:10.1088/1755-1315/23/1/011001

Intimate links exist between cities and questions of climate chang

Unified Satellite Image Fusion for
Sustainable Urban Environment
Bo Huang
Department of Geography and Resource Management &
Institute of Space and Earth Information Science
The Chinese University of Hong Kong

Peer review statement

All papers published in this volume of *IOP Conference Series: Earth and Environmental Science* have been peer reviewed through processes administered by the proceedings Editors. Reviews were conducted by expert referees to the professional and scientific standards expected of a proceedings journal published by IOP Publishing.

IOP Conf. Series: Earth and Environmental Science **23** (2015) 012001 doi:10.1088/1755-1315/23/1/012001

Soliton solutions for higher grading affine Toda from quantum algebras

Alexander Zuevsky

Institute of Mathematics, Academy of Sciences of the Czech Republic, Prague

E-mail: `zuevsky@yahoo.com`

Abstract. Based on quantum Lie algebras associated to affine Lie algebras we construct special soliton-like solutions for higher grading affine Toda models.

1. Introduction

The group–theoretical (Lie-algebraic) method [6, 7] of construction and integration of two-dimensional exactly solvable models is one of the most powerful. Using the representation theory for conformal and affine Toda models [9, 7] one is able to reconstruct the general solutions to systems as well as to derive special, in particular, solitonic solutions both in classical and quantum regions. The zero curvature condition applied to elements of connections constructed as sums of products of arbitrary functions with generators of (higher) grading subspaces [5] of a Lie algebra results in systems of equations [3].

Previously, general solution to the higher grading affine Toda equation were obtained in [3]. In [9] group-theoretical specialization for the lower grading subspace were formulated, and classical soliton solutions obtained in frames of algebraic approach to integrable models. In [10] we proposed quantum Lie algebra solutions generating solitons based on further algebraic properties of quantum groups.

In this paper we consider the case of affine Toda models constructed using algebraic elements that belong to grading subspaces up to some number l. The group–theoretical approach allows to extract special solitonic solutions from general ones. The lower grading case for the quantized version of the affine Toda models (the sine–Gordon equation) involving the quantum Lie algebra [1] $(sl_2)_q$ was considered in [10]. Here we extended that construction for a higher grading subspace case. Exactly solvable dynamical systems appearing in this the higher grading formulation are generalizations of lower ($l = 1$) cases, and possess more complicated and deep properties describing more complicated physical situations. It is important to take into account next terms in more general equations in order to understand interrelations of ordinary solutions (to equations associated to lower grading algebraic generators) with solutions corresponding to higher graiding.

2. Higher grading affine Toda system

In this and the next sections we recall [3] the affine Toda system consrtuction. Consider a two dimensional manifold $\mathcal{M}$ with local coordinates $z_{\pm}$. Up to a gauge transformation, $(1, 0)$-component lying in (see subsection 7.1 of Appendix) $\bigoplus_{n=0}^{l} \widehat{\mathcal{G}}_{+n}$ and $(0, 1)$-component in

$\bigoplus_{n=0}^{l} \widehat{\mathcal{G}}_{-n}$ of a flat connection $\mathcal{A}$ in the trivial holomorphic principal fibre bundle $\mathcal{M} \times \widehat{G} \longrightarrow \mathcal{M}$ ($l > 0$ is fixed integer) satisfy the zero curvature condition

$$\partial_+ A_- - \partial_- A_+ + [A_+ \,,\, A_-] = 0. \tag{1}$$

The components $A_\pm$ are the following (we keep notations of [3])

$$A_+ = -B\,F^+\,B^{-1}, \qquad A_- = -\partial_- B\,B^{-1} + F^-. \tag{2}$$

Here B is a mapping $\mathcal{M} \longrightarrow \widehat{G}_0$ ($\widehat{G}_0$ is a group with the Lie algebra $\widehat{\mathcal{G}}_0$) and $F^\pm$ $(1 \le m \le l-1)$ are mappings to $\bigoplus_{n=1}^{l} \widehat{\mathcal{G}}_{\pm n}$

$$F^\pm = E_{\pm l} + \sum_{m=1}^{l-1} F^\pm_m, \tag{3}$$

where $E_{\pm l}$ are some fixed elements of $\widehat{\mathcal{G}}_{\pm l}$ and $F^\pm_m \in \widehat{\mathcal{G}}_{\pm m}$, $(1 \le m \le l-1)$. Substituting (2) into (1) one arrives at the equations of motion

$$\partial_+\left(\partial_- B\,B^{-1}\right) = [E_{-l}\,,\,B\,E_l\,B^{-1}] + \sum_{n=1}^{l-1}[F^-_n\,,\,B\,F^+_n\,B^{-1}], \tag{4}$$

$$\partial_- F^+_m = [E_l\,,\,B^{-1}\,F^-_{l-m}\,B] + \sum_{n=1}^{l-m-1}[F^+_{n+m}\,,\,B^{-1}\,F^-_n\,B], \tag{5}$$

$$\partial_+ F^-_m = -[E_{-l}\,,\,B\,F^+_{l-m}\,B^{-1}] - \sum_{n=1}^{l-m-1}[F^-_{n+m}\,,\,B\,F^+_n\,B^{-1}]. \tag{6}$$

Since $Q_{\mathbf{s}}$, $C \in \widehat{\mathcal{G}}_0$ then B can be parameterized as $B = b\,e^{\eta\,Q_{\mathbf{s}}}\,e^{\nu\,C}$ where b is a mapping to G_0, the subgroup of $\widehat{G}_0$ generated by all elements of $\widehat{\mathcal{G}}_0$ other than $Q_{\mathbf{s}}$ and C. Substituting B into the equations of motion (4–6) one has

$$\partial_+\left(\partial_- b b^{-1}\right) + \partial_+\partial_-\,\nu\,C = e^{l\eta}[E_{-l}\,,\,b\,E_l\,b^{-1}] + \sum_{n=1}^{l-1} e^{n\eta}\,[F^-_n\,,\,b\,F^+_n b^{-1}], \tag{7}$$

$$\partial_- F^+_m = e^{(l-m)\eta}\,[E_l\,,\,b^{-1}\,F^-_{l-m}\,b] + \sum_{n=1}^{l-m-1} e^{n\eta}[F^+_{m+n}\,,\,b^{-1}\,F^-_n\,b], \tag{8}$$

$$\partial_+ F^-_m = -e^{(l-m)\eta}\,[E_{-l}\,,\,b\,F^+_{l-m}\,b^{-1}] - \sum_{n=1}^{l-m-1} e^{n\eta}[F^-_{m+n}\,,\,b\,F^+_n\,b^{-1}], \tag{9}$$

$$\partial_+\partial_-\,\eta\,Q_{\mathbf{s}} = 0. \tag{10}$$

Now consider the case $l = 1$. Let us parameterize the element B in principal grading of $\widehat{\mathcal{G}}$ and take $b = e^{(\phi(x_+,x_-),H^0)}$, $\gamma_0^\pm(x_\pm) = e^{(\phi^\pm(x_\pm),H^0)}$, where $(\phi, H^0) \equiv \sum_{i=1}^{r} \phi_i H^0_i$. From the equations (7–10) for an infinite dimensional Lie algebra $\widehat{\mathcal{G}}$ in the principal grading we obtain the affine Toda field theory systems of equations

$$\partial_+\partial_-\phi + \tfrac{4\mu}{\beta}\sum_{i=1}^{r}\left(m_i \tfrac{\alpha_i}{\alpha_i^2}\exp(\beta\alpha_i\cdot\phi) - \tfrac{\psi}{2}\exp(-\beta\cdot\phi)\right) = 0. \tag{11}$$

The formal general solution to the above equation was introduced in [9]:

$$e^{-\beta\lambda_i\cdot\phi} = e^{-\beta\lambda_i\cdot\phi_0}\frac{{}^{(1)}\langle\lambda_i|(\gamma_0^+)^{-1}\mu_+^{-1}(z_+)\mu_-(z_-)(\gamma_0^-)|\lambda_i\rangle^{(1)}}{\left({}^{(1)}\langle\lambda_0|(\gamma_0^+)^{-1}\mu_+^{-1}(z_+)\mu_-(z_-)(\gamma_0^-)|\lambda_0\rangle^{(1)}\right)^{m_j}} = e^{-\beta\lambda_i\cdot\phi_0}\frac{(1)\langle\lambda_i|B^{-1}|\lambda_i\rangle^{(1)}}{\left({}^{(1)}\langle\lambda_0|B^{-1}|\lambda_0\rangle^{(1)}\right)^{m_j}}, \tag{12}$$

The fact that (12) is indeed a solution to (11) may be checked by using the representation theory of $\widehat{\mathcal{G}}$. A map $g:\mathcal{M}\longrightarrow G$ appearing in the gradient form of the flat connection $A_\pm = g^{-1}\partial_\pm g$, may be factorized (according to the Lie algebra decomposition $\mathcal{G}=\mathcal{G}_-\oplus\mathcal{G}_0\oplus\mathcal{G}_+$) by the modified Gauss decomposition $g=\mu_-\nu_+\gamma_{0-}$ or $g=\mu_+\nu_-\gamma_{0+}$ with maps $\gamma_{0\pm}:\mathcal{M}\longrightarrow G_0$, $\mu_\pm,\nu_\pm:\mathcal{M}\longrightarrow G_\pm$. The grading condition provides the holomorphic property of $\mu_\pm$, i.e., they satisfy the initial value problem

$$\partial_\pm\mu_\pm(z_\pm)=\mu_\pm(z_\pm)\tilde{\mathcal{E}}_\pm(z_\pm), \tag{13}$$

$$\tilde{\mathcal{E}}_\pm(z_\pm)=\sum_{m=1}^{M}\tilde{\mathcal{E}}^\pm_m(\Phi^\pm),\qquad \tilde{\mathcal{E}}^\pm_m(\Phi^\pm)=\sum_{\alpha\in\Delta^+_m}\Phi^{\pm m}_\alpha(z_\pm)X_{\pm\alpha}, \tag{14}$$

with arbitrary functions $\Phi^{\pm m}_\alpha(z_\pm)$ determining the general solution to the system. Note that the summations in (14) are performed over the set of positive roots Δ^+_m of $\mathcal{G}=\sum_{m\in\mathbb{Z}}\mathcal{G}_m$ in the subspace $\mathcal{G}_m$. Using notations of [3], $\widehat{g}_0\equiv\gamma_{0+}\gamma_{0-}^{-1}$, and $\gamma_0^\pm(x_\pm):\mathcal{M}\longrightarrow G_0$ are arbitrary mappings. The element B is parameterized as $B^{-1}=(\gamma_0^+(x_+))^{-1}\widehat{g}_0\gamma_0^-(x_-)$.

3. Generalization of the affine Toda equations

Consider for example the case of $\widehat{sl}_2$ in principal grading (see subsection 7.1 of Appendix). The mapping B is parameterised as

$$B=e^{\varphi\,H^0}\,e^{\tilde{\nu}\,C}\,e^{\eta\,Q}=e^{\varphi\,\tilde{H}^0}\,e^{\nu\,C}\,e^{\eta\,Q}, \tag{15}$$

where $\tilde{H}^0=H^0-\frac{1}{2}C$ is Cartan element, and $\tilde{\nu}=\nu-\frac{1}{2}\varphi$. The case $l=3$ delivers a generalization of (11). In this case $E_3=a_+E^1_++a_-E^2_-$, $E_{-3}=a_+E^{-2}_++a_-E^{-1}_-$, [3] and therefore $[E_3\,,\,E_{-3}]=3a_+a_-C\equiv\beta C$. Introduce the notations

$$E_1=a_+E^0_++a_-E^1_-,\ E_{-1}=a_+E^{-1}_++a_-E^0_-,\ f_1=a_+E^0_+-a_-E^1_-,\ f_{-1}=a_+E^{-1}_+-a_-E^0_-,$$
$$f_2=-\sqrt{a_+a_-}\,H^1,\quad f_{-2}=-\sqrt{a_+a_-}\,H^{-1}.$$

The fields of the model are in (15) as well as matter fields $\psi^i_{R/L}$, $i=1,\,2$, $\chi_\pm$ defined as

$$F_1^+=\psi^2_Rf_1+\chi_+E_1,\qquad F_2^+=\psi^1_Rf_2,\quad F_1^-=\psi^1_Lf_{-1}+\chi_-E_{-1},\qquad F_2^-=\psi^2_Lf_{-2},$$

$\psi^i_{R/L}$, $i=1,2$, are components of two the Dirac fields ψ^i. It follows then that

$$[E_3,[E_{-3},E_{\pm1}]]=0,\ [E_3,[E_{-3},\tilde{H}^0]]=4a_+a_-\tilde{H}^0,\ [E_3,[E_{-3},f_i]]=4a_+a_-f_i,\ i=\pm1,\pm2.$$

Using (7–10), the equations of motion are

$$\begin{aligned}
\partial_+\partial_-\varphi &= -a_+a_-\left(\left(e^{2\varphi}-e^{-2\varphi}\right)\left(e^{3\eta}-e^{\eta}\left(\psi_L^1\psi_R^2-\chi_+\chi_-\right)\right)\right.\\
&- \left. e^{\eta}\left(e^{2\varphi}+e^{-2\varphi}\right)\left(\psi_L^1\chi_+-\psi_R^2\chi_-\right)\right), \qquad (16)\\
\partial_+\partial_-\nu &= -a_+a_-\left(\left(e^{2\varphi}+e^{-2\varphi}\right)\left(\frac{3}{2}e^{3\eta}-\frac{1}{2}e^{\eta}\left(\psi_L^1\psi_R^2-\chi_+\chi_-\right)\right)\right.\\
&- \left.\frac{1}{2}e^{\eta}\left(e^{2\varphi}-e^{-2\varphi}\right)\left(\psi_L^1\chi_+-\psi_R^2\chi_-\right)+2e^{2\eta}\psi_L^2\psi_R^1\right), \qquad (17)\\
\partial_-\psi_R^1 &= \sqrt{a_+a_-}e^{\eta}\left(\psi_L^1\left(e^{2\varphi}+e^{-2\varphi}\right)-\chi_-\left(e^{2\varphi}-e^{-2\varphi}\right)\right), \qquad (18)\\
\partial_+\psi_L^1 &= 2\sqrt{a_+a_-}\left(-e^{2\eta}\psi_R^1+\frac{1}{2}e^{\eta}\psi_L^2\left(\psi_R^2\left(e^{2\varphi}-e^{-2\varphi}\right)+\chi_+\left(e^{2\varphi}+e^{-2\varphi}\right)\right)\right), \qquad (19)\\
\partial_-\psi_R^2 &= 2\sqrt{a_+a_-}\left(e^{2\eta}\psi_L^2+\frac{1}{2}e^{\eta}\psi_R^1\left(\psi_L^1\left(e^{2\varphi}-e^{-2\varphi}\right)-\chi_-\left(e^{2\varphi}+e^{-2\varphi}\right)\right)\right), \qquad (20)\\
\partial_+\psi_L^2 &= \sqrt{a_+a_-}e^{\eta}\left(-\psi_R^2\left(e^{2\varphi}+e^{-2\varphi}\right)-\chi_+\left(e^{2\varphi}-e^{-2\varphi}\right)\right), \qquad (21)\\
\partial_-\chi_+ &= \sqrt{a_+a_-}e^{\eta}\psi_R^1\left(\chi_-\left(e^{2\varphi}-e^{-2\varphi}\right)-\psi_L^1\left(e^{2\varphi}+e^{-2\varphi}\right)\right), \qquad (22)\\
\partial_+\chi_- &= \sqrt{a_+a_-}e^{\eta}\psi_L^2\left(\chi_+\left(e^{2\varphi}-e^{-2\varphi}\right)+\psi_R^2\left(e^{2\varphi}+e^{-2\varphi}\right)\right), \qquad (23)\\
\partial_+\partial_-\eta &= 0. \qquad (24)
\end{aligned}$$

The general solutions to the matter fields $F_i^{\pm}$ may be written in the following form. For $m=1$ in (7–10) one has [3]

$$\langle i|F_1^+|i;i\rangle = f_i^+ = e^{\sum_{l=0}^{r} k_{il}(\phi^- - \phi)_l} e^{\nu_0}\partial_+\left(\langle i|\mu_+^{-1}\mu_-|i;i\rangle\frac{\langle 0|\mu_+^{-1}\mu_-|0\rangle^{m_i}}{\langle i|\mu_+^{-1}\mu_-|i\rangle}\right). \qquad (25)$$

Here $|i;i\rangle$ denotes an element of the Verma module which is result of the action of the lowering generator on the highest state vector.

4. Solitonic solutions from general solutions

In [9] it was shown how to extract solitonic solutions from the formal general solutions of the affine Toda field equations. Let's take $\gamma_0^{\pm}=1$ in (12) to be a constant function. Then the mappings $\mu_{\pm}$ are $\mu_{\pm}=\mu_{\pm}^0 e^{z_{\pm}\mathcal{E}_{\pm}}$ with $\mu_{\pm}^0$ being some fixed mappings independent of $z_{\pm}$. Next take $\tilde{\mathcal{E}}_{\pm}$ in (14) as $\mathcal{E}_{\pm}\equiv E_{\pm l}+\sum_{N=1}^{l-1}c_N^{\pm}E_{\pm N}$ where $E_{\pm}$ are elements of a Heisenberg subalgebra of $\widehat{\mathcal{G}}$, namely $[\mathcal{E}_+,\mathcal{E}_-]=\Omega C$. One can consider principal of homogenious Heisenberg subalgebras for that purpose. In this paper we only deel with the principal case while the homogeneous case will be discussed elsewhere. Thus, we arrive at a special solution to (11)

$$e^{-\beta\lambda_i\cdot\phi} = e^{-\beta\lambda_i\cdot\phi_0}\frac{{}^{(1)}\langle\lambda_i|e^{x_{\pm}\mathcal{E}_{\pm}}\mu^0 e^{x_{\pm}\mathcal{E}_{\pm}}|\lambda_i\rangle^{(1)}}{\left({}^{(1)}\langle\lambda_0|e^{x_{\pm}\mathcal{E}_{\pm}}\mu^0 e^{x_{\pm}\mathcal{E}_{\pm}}|\lambda_0\rangle^{(1)}\right)^{m_j}}. \qquad (26)$$

In order to compute these solutions explicitly we have to remove $\mathcal{E}_{\pm}$-dependence from (26) moving $\mathcal{E}_+$ to the right and $\mathcal{E}_-$ to the left . Then we should find such a $\mu_0=\prod_{i=1}^{N}e^{\mathcal{V}_i}$ so that $\mathcal{V}_i$ would be eigenvectors with respect to the adjoint action of $\mathcal{E}_{\pm}$, i.e., $[\mathcal{E}_{\pm},\mathcal{V}_i]=\omega_{\pm}^{(i)}\mathcal{V}_i$. Then it turns out [9] that resulting expressions provide us with solitonic solutions to the equations under considerations while parameters $\omega_{\pm}^{(i)}$ characterize solitons.

In the next sections we will explain the affinization $(sl_2)_q^t$ of the quantum Lie algebra $(sl_2)_q$ [1] and then apply those formulations to the construction of q-deformed solutions to the quantised [8] affine Toda equations in the specific example of the higher grading the sine–Gordon equation.

5. Quantum Lie algebra

5.1. *Universal enveloping algebra $U_q(sl_2)$*

The complex Lie algebra sl_2 is generated by three elements X_+, X_- and H satisfying

$$HX^{\pm} - X^{\pm}H = \pm 2X^{\pm}, \qquad X^+X^- - X^-X^+ = H.$$

The Lie bracket is skew-symmetric and satisfies Jacobi identity. The quantized enveloping algebra $U_q(sl_2)$ [2], [4] is an associative algebra generated by X^+, X^-, h with q-deformed commutator relations

$$X^+X^- - X^-X^+ = \tfrac{q^H - q^{-H}}{q-q^{-1}}, \qquad HX^{\pm} - X^{\pm}H = \pm 2X^{\pm},$$

It possesses a Hopf algebra structure with deformed adjoint action

$$(ad_{X^{\pm}})_q a = X^{\pm} a q^{H/2} - q^{\mp 1} q^{H/2} a X^{\pm}, \qquad (ad_H)_q a = Ha - aH, \tag{27}$$

(for all $a \in U_q(sl_2)$).

5.2. *Quantum algebra $(sl_2)_q$*

In [1] new generators in the quantized universal enveloping algebra $U_q(\widehat{sl_2})$ were introduced as

$$X_h^{\pm} = \sqrt{\tfrac{2}{q+q^{-1}}}\, q^{-H/2} X^{\pm}, \qquad H_h = \tfrac{2}{q+q^{-1}} \left(qX^+X^- - q^{-1}X^-X^+\right). \tag{28}$$

It forms a three-dimensional subspace $(sl_2)_q = X_h^+, X_h^-, H_h$ in $U_q(sl_2)$, closed under the quantum Lie bracket

$$[a,b]_h = (ad_a)_q b, \tag{29}$$

a, $b \in (sl_2)_q$. These generators possesses the following commutation relations

$$\begin{aligned} &[H_h, X_h^{\pm}]_h = \pm 2q^{\pm 1} X_h^{\pm},\ [X_h^{\pm}, H_h]_h = \mp 2q^{\mp 1} X_h^{\pm},\ [X_h^+, X_h^-]_h = H_h, \\ &\left[X_h^-, X_h^+\right]_h = -H_h\ [H_h, H_h]_h = 2(q - q^{-1})H_h,\ [X_h^{\pm}, X_h^{\pm}]_h = 0. \end{aligned} \tag{30}$$

We should mention that $(sl_2)_q$ is not a Lie algebra in the standard sense. The generators (28) do not satisfy Jacobi identity and q-deformed Lie bracket is not skew-symmetric. Although q-analog of Jacobi identity for $(sl_2)_q$ is still absent, nevertheless (29) is q-skew-symmetric, [1]. Under q-skew-symmetry we mean a symmetry under q-conjugation, (that we will denote by tilde) an automophism of $(sl_2)_q$ defined by $q \mapsto \frac{1}{q}$. Then for any element of $(sl_2)_q$ $(aX_h^+ + bX_h^- + cH_h)\tilde{} = \tilde{a}X_h^+ + \tilde{b}X_h^- + \tilde{c}H_h$, ($a$, b, $c \in C$) and q-deformed Lie bracket satisfies $\widetilde{[a,b]_h} = [\tilde{b}, \tilde{a}]_h$.

5.3. *Affinization of the $(sl_2)_q$*

In [10] an affinized version, the principal Heisenberg subalgebra, an eigenvalue vertex operator, and representations of the quantum algebra $(sl_2)_q$ were considered. In this section and subsection 7.2 of Appendix we recall these constructions and then apply them in section 6 to the construction of solutions for the higher grading the sine–Gordon equation. Let $\mathcal{G} = (sl_2)_q$ and $\mathcal{L} = C[t, t^{-1}]$ be the algebra of Lourant polynomials in t and let $\mathcal{L}(\mathcal{G}) = \mathcal{L} \otimes_C \mathcal{G}$. Introduce a complex vector space $(sl_2)_q^t : \widetilde{\mathcal{L}}(\mathcal{G}) = \mathcal{L}(\mathcal{G}) \oplus CK \oplus Cd$. This is a loop algebra $\mathcal{L}(\mathcal{G})$ completed with the derivation d (acting as $t\frac{d}{dt}$ in $\mathcal{L}$ and trivially on K) extended by a one-dimensional center K corresponding to C-valued q-deformed 2-cocycle on $\mathcal{L}(\mathcal{G})$ $\Psi_q(a,b) = (x|y)_h \Phi(P,Q)$,

$\Phi(P,Q) = Res_t \frac{dP}{dt}Q$. Here $(x|y)_h$ is a quantum bilinear form on $(sl_2)_q$ [1] and P, Q are polynoms in t. We define q-deformed Lie bracket in this algebra as

$$\begin{aligned}&[t^m \otimes x \oplus \omega K \oplus \nu d, \quad t^n \otimes y \oplus \omega_1 K \oplus \nu_1 d]_h = \\ &(t^{m+n} \otimes [x,y]_h + \nu n t^n \otimes y - n\nu_1 m t^m \otimes x) \oplus m\delta_{m+n,0}(x|y)_h K\end{aligned} \tag{31}$$

where $x,\ y \in \mathcal{G},\ \nu,\ \omega,\ \nu_1,\ \omega_1 \in C$.

Now we introduce generators that constitute an affinization of quantum algebra $(sl_2)_q$

$$H_1 = 1 \otimes H_h,\ H_0 = 1 \otimes (K - H_h),\ e_1 = 1 \otimes X_h^+,\ e_0 = t \otimes X_h^-,\ f_1 = 1 \otimes X_h^-,\ f_0 = t^{-1} \otimes X_h^+.$$

Then one can calculate the adjoint action

$$\begin{aligned}
&[H_0, e_0]_h = 2e_0 q^{-1}, [H_0, f_0]_h = -2f_0 q^{-1}, [H_0, e_1]_h = -2e_1 q, [H_0, f_1]_h = 2f_1 q^{-1}, \\
&[H_1, e_1]_h = 2e_1 q, [H_1, f_1]_h = -2f_1 q^{-1}, [H_1, e_0]_h = 2e_0 q^{-1}, [H_1, f_0]_h = 2q f_0, \\
&[H_0, H_1]_h = -2(q - q^{-1})H_1, [H_0, H_0]_h = 2(q - q^{-1})H_1, [H_1, H_1]_h = 2(q - q^{-1})H_1, \\
&[H_1, H_0]_h = -2(q - q^{-1})H_1, [e_0, f_0]_h = H_0, [e_1, f_1]_h = H_1, [f_0, e_0]_h = -H_0, [f_1, e_1]_h = -H_1, \\
&[e_0, H_0]_h = -2qe_0, [e_0, H_1]_h = 2qe_0, [e_1, H_0]_h = 2e_1 q, [e_1, H_1]_h = -2e_1 q, \\
&[f_0, H_0]_h = 2q^{-1} f_0, [f_0, H_1]_h = -2q^{-1} f_0, [f_1, H_1]_h = 2q f_1, [f_1, H_0]_h = -2q f_1.
\end{aligned}$$

5.4. *Heisenberg subalgebra and eigenvector of the q-deformed adjoint action*

The key point in the construction of solitonic solution to the affine Toda equation is the existence of eigenvectors with respect to elements of the Heisenberg subalgebra of the underlying affine algebra. Let

$${}_qE_{+1} = 1 \otimes X_h^+ + t \otimes X_h^-, \qquad {}_qE_{-1} = 1 \otimes X_h^- + t^{-1} \otimes X_h^+$$

so that the following series

$$\begin{aligned}&A_{2m} = (q + q^{-1})^m t^m \otimes (-H_h),\ A_{2m+1} = (q + q^{-1})^m t^m \otimes q^{-1} X_h^+ - t^{m+1} \otimes q X_h^- + \left(X_h^- | X_h^+\right)_h, \\ &\mathcal{F}_q = \sum_{k=-\infty}^{+\infty} \zeta^k A_k,\end{aligned} \tag{32}$$

is an eigenvector of ${}_qE_{\pm 1}$ with respect to the quantum affine Lie bracket (31). Using $\mathcal{F}_q$ we can find q-deformation of the solitonic solution (26) corresponding to the affinization of the quantum algebra $(sl_2)_q$.

6. Solution to the affine Toda equations corresponding to $(sl_2)_q^t$

In this section we will use results of the section 5. We propose q-deformed solitonic solutions to the affine Toda equations associated with the affine Lie algebra $\widehat{sl}_2$, i.e., the sine–Gordon equation. Those solutions will be constructed by means of the Heisenberg subalgebra operators and the eigenvector $\mathcal{F}$ involved in the exponent in the mapping μ_0. Using properties of the highest weight representation of the affinized quantum algebra $(sl_2)_q$ one can check (similar to [9], [8]) that (12) is a solution to the affine Toda equation (11) when we replace its ingredients by their quantum group counterparts. A consideration in the case of the sine–Gordon equation is given in [9].

First consider the case of $l = 1$. Starting from the general solution (12) to the affine Toda equations for $\widehat{sl}_2$ we take $\gamma_0^{\pm} = 1$ and substitute the vectors of the fundamental highest weight representation of $\widehat{sl}_2$ by the fundamental highest weight representation vectors of $(sl_2)_q^t$

$$e^{-\beta\lambda_i \cdot \phi} = e^{-\beta\lambda_i \cdot \phi_0} \frac{{}_q^{(1)}\langle \lambda_i | \mu_+^{-1}(z_+)\mu_-(z_-)|\lambda_i\rangle_q^{(1)}}{\left({}_q^{(1)}\langle \lambda_0 | \mu_+^{-1}(z_+)\mu_-(z_-)|\lambda_0\rangle_q^{(1)}\right)^{m_j}}, \tag{33}$$

IOP Conf. Series: Earth and Environmental Science **23** (2015) 012001 doi:10.1088/1755-1315/23/1/012001

where $|i\rangle_q^{(1)} = |\lambda_i\rangle_q^{(1)} = v_0^{(i)}$ denotes the highest vector in the $i-$th fundamental representation of $(sl_2)_q^t$. Taking $\mathcal{E}_\pm = {}_qE_{\pm 1}$ in the solitonic solution and inserting exponential of the eigen operator $\mathcal{F}_q$ as a group element, we obtain [10]

$$e^{-\beta\lambda_j\cdot\phi} = e^{-\beta\lambda_j\cdot\phi_0}\frac{{}_q^{(1)}\langle j|e^{-\mathcal{E}_+z_+}e^{Q\mathcal{F}_q}e^{\mathcal{E}_-z_-}|j\rangle_q^{(1)}}{\left({}_q^{(1)}\langle 0|e^{\mathcal{E}_+z_+}e^{Q\mathcal{F}_q}e^{\mathcal{E}_-z_-}|0\rangle_q^{(1)}\right)^{m_j}}. \tag{34}$$

Then it follows that

$$e^{-\beta\lambda_j\cdot\phi} = e^{-\beta\lambda_j\cdot\phi_0}\frac{{}_q^{(1)}\langle j|\exp\left(Qe^{-2z_+\zeta-2\frac{1}{t}z_-\zeta^{-1}}\mathcal{F}_q\right)\exp(-z_+z_-)|j\rangle_q^{(1)}}{\left({}_q^{(1)}\langle 0|\exp\left(Qe^{-2z_+\zeta-2\frac{1}{t}z_-\zeta^{-1}}\mathcal{F}_q\right)\exp(-z_+z_-)|0\rangle_q^{(1)}\right)^{m_j}}. \tag{35}$$

So for $j=0$

$$e^{-\beta\lambda_1\cdot\phi} = e^{-\beta\lambda_1\cdot\phi_0}, \tag{36}$$

and for $j=1$ this gives

$$e^{-\beta\lambda_1\cdot\phi} = e^{-\beta\lambda_1\cdot\phi_0}\exp\left(-Qe^{-2z_+\zeta-2\frac{1}{t}z_-\zeta^{-1}}\sum_{m=-\infty}^{+\infty}\zeta^{2m}t^m(q+q^{-1})^m\right). \tag{37}$$

6.1. The case $l=3$

The affine Toda equation system (11) is a particular example of more general construction described in the first section. Namely equations (11) correspond to $l=1$ grading subspace. The case $l=3$ leads us to a generalization of (11). Using the general form of solution to the affine Toda equation for $l>1$ (12) we can form a class of q-deformed solutions constructed with by means of the eigenvector $\mathcal{F}_q$ (32). We take in (12)

$$\mathcal{E}_\pm = {}_qE_{\pm 1} + {}_qE_{\pm 3}, \tag{38}$$

where ${}_qE_{+3} = t\otimes X_h^+ + t^2\otimes X_h^-$, ${}_qE_{-3} = t^{-1}\otimes X_h^- + t^{-2}\otimes X_h^+$. Then we arrive at a solution for Toda type field ϕ

$$e^{-\beta\lambda_1\cdot\phi} = e^{-\beta\lambda_1\cdot\phi_0}\exp\left(-Qe^{-2z_+(1+t)\zeta-2\frac{1}{t}z_-(1+\frac{1}{t})\zeta^{-1}}\sum_{m=-\infty}^{+\infty}\zeta^{2m}t^m(q+q^{-1})^m\right). \tag{39}$$

Now we want to find q-deformed solitonic solutions for matter fields $F_m^\pm$ just in the same way as we did it for Toda type fields. We substitute elements (38) to (25) and take the state vectors from (42) from the basis of the fundamental highest weight representation of $(sl_2)_q^t$. Finally the q-solitonic solutions are

$$\begin{aligned}
f_1^+ &= \exp\left(\sum_{l=0}^{1}k_{1l}(\phi^- - \phi)_l\right)e^{\nu_0}\partial_+\left({}_q\langle 1|\exp\left(Qe^{-2z_+(1+t)\zeta-2\frac{1}{t}z_-(1+\frac{1}{t})\zeta^{-1}}\sum_{m=-\infty}^{+\infty}\zeta^{2m}t^m(q+q^{-1})^m\right)\right.\\
&\times\sum_{n=0}^{+\infty}\frac{1}{n!}(Qe^{-2z_+(1+t)\zeta-2\frac{1}{t}z_-(1+\frac{1}{t})\zeta^{-1}})^n\, q^{-1}q^{-\lambda^{(1)}}\\
&\times\left(\sum_{m=-\infty}^{+\infty}\zeta^{2m+1}t^m(q+q^{-1})^m\right)^n\left.\left((1-2qq^{-(\lambda^{(1)}-2)})^n|1;1\rangle_q + \sum_{k=0}^{n-1}(1-2qq^{-(\lambda^{(1)}-2)})^k|1\rangle_q\right)\right)\\
&= e^{\sum_{l=0}^{1}k_{1l}(\phi^- - \phi)_l}e^{\nu_0}\partial_+\left(\exp\left(2Qe^{-2z_+(1+t)\zeta-2\frac{1}{t}z_-(1+\frac{1}{t})\zeta^{-1}}\sum_{m=-\infty}^{+\infty}\zeta^{2m}t^m(q+q^{-1})^m\right)q^{-1}q^{-\lambda^{(1)}}\right.\\
&\times\left.\left(\exp(1-2qq^{-(\lambda^{(1)}-2)})|1;1\rangle_q + \frac{\exp\left((1-2qq^{-(\lambda^{(1)}-2)})\right)-1}{(1-2qq^{-(\lambda^{(1)}-2)})-1}|1\rangle_q\right)\right),
\end{aligned}$$

$$f_0^+ = \exp\left(\sum_{l=0}^{1} k_{1l}(\phi^- - \phi)_l\right) e^{\nu_0}\partial_+ \left(\exp\left(Qe^{-2z_+(1+t)\zeta - 2\frac{1}{t}z_-(1+\frac{1}{t})\zeta^{-1}} \sum_{m=-\infty}^{+\infty} \zeta^{2m}t^m(q+q^{-1})^m\right)\right.$$

$$\left. \times q^{-1}q^{-\lambda^{(0)}} \left(\exp(1-2q^{-1}q^{(2+\lambda_0^{(0)})})|0;0\rangle_q + \frac{\exp\left((1-2q^{-1}q^{(2+\lambda_0^{(0)})})\right)-1}{(1-2q^{-1}q^{(2+\lambda_0^{(0)})})-1}|0\rangle_q\right)\right).$$

Similar relations can be obtain all higher cases of l elsewhere.

7. Appendix

7.1. Affine Kac–Moody algebras

Here we recall facts about affine Kac–Moody algebras [5], [3]. Consider an untwisted affine Kac-Moody algebra $\widehat{\mathcal{G}}$ endowed with an integral grading $\widehat{\mathcal{G}} = \bigoplus_{n\in\mathbb{Z}} \widehat{\mathcal{G}}_n$, and denote $\widehat{\mathcal{G}}_\pm = \bigoplus_{n>0} \widehat{\mathcal{G}}_{\pm n}$. By an affine Lie algebra we mean a loop algebra corresponding to a finite dimensional simple Lie algebra $\mathcal{G}$ of rank r, extended by the center C and the derivation D. According to [5], integral gradings of $\widehat{\mathcal{G}}$ are labelled by a set of co-prime integers $\mathbf{s} = (s_0, s_1, \dots s_r)$, and the grading operators are given by

$$Q_{\mathbf{s}} \equiv H_{\mathbf{s}} + N_{\mathbf{s}}\, D - \frac{1}{2N_{\mathbf{s}}}\,\mathrm{Tr}\,(H_{\mathbf{s}})^2\, C\,. \tag{40}$$

Here $H_{\mathbf{s}} \equiv \sum_{a=1}^r s_a\lambda_a^v \cdot H^0$, $N_{\mathbf{s}} \equiv \sum_{i=0}^r s_i m_i^\psi$, $\psi = \sum_{a=1}^r m_a^\psi \alpha_a\,,\quad m_0^\psi = 1$. H^0 is an element of Cartan subalgebra of $\mathcal{G}$; α_a, $a = 1, 2, \dots r$, are its simple roots; ψ is its maximal root; m_a^ψ the integers in expansion $\psi = \sum_{a=1}^r m_a^\psi \alpha_a$; and λ_a^v are the fundamental co–weights satisfying the relation $\alpha_a \cdot \lambda_b^v = \delta_{ab}$.

The principal grading operator Q_{ppal} is given by (40) where $N_{\mathbf{s}} = h$ is Coxeter number. Therefore $\widehat{\mathcal{G}}_0 = \{H_a^0, a = 1, 2, \dots r; C; Q_{\mathrm{ppal}}\}$, $\widehat{\mathcal{G}}_m = \{E^0_{\alpha^{(m)}}, E^1_{-\alpha^{(h-m)}}\}$, $\widehat{\mathcal{G}}_{-m} = \{E^0_{-\alpha^{(m)}}, E^{-1}_{\alpha^{(h-m)}}\}$ where $0 < m < h$, and $\alpha^{(m)}$ are positive roots of height m. The element B is parameterized as $B = e^{\varphi\cdot\tilde{H}^0}\, e^{\nu\, C}\, e^{\eta Q_{\mathrm{ppal}}} = e^{\varphi\cdot H^0}\, e^{\tilde{\nu}\, C}\, e^{\eta Q_{\mathrm{ppal}}}$, where $\tilde{H}^0$ was defined in [3] as $\widetilde{H}_a^0 = H_a^0 - \frac{1}{N_{\mathbf{s}}}\,\mathrm{Tr}\left(H_{\mathbf{s}}\, H_a^0\right) C = H_a^0 - \frac{2}{\alpha_a^2}\frac{s_a}{N_{\mathbf{s}}} C$, and $\tilde{\nu} = \nu - \frac{2}{h}\widehat{\delta}\cdot\varphi$, with $\widehat{\delta} = \sum_{a=1}^r \frac{\lambda_a}{\alpha_a^2}$, and λ_a being the fundamental weights of $\mathcal{G}$. Let us denote by H^n, $E_\pm^n$, D, C the Chevalley basis generators of $\widehat{sl}_2$. The commutation relations are

$$\begin{aligned} [H^m\,,\, H^n] = 2\,m\,C\,\delta_{m+n,0}, \quad [H^m\,,\, E_\pm^n] = \pm 2\, E_\pm^{m+n}, \\ [E_+^m\,,\, E_-^n] = H^{m+n} + m\,C\,\delta_{m+n,0}, \quad [D\,,\, T^m] = m\,T^m\,, \quad T^m \equiv H^m, E_\pm^m. \end{aligned} \tag{41}$$

The grading operator for the principal grading ($\mathbf{s} = (1,1)$) is $Q \equiv \frac{1}{2}H^0 + 2D$. Then the eigensubspaces are $\widehat{\mathcal{G}}_0 = \{H^0, C, Q\}$, $\widehat{\mathcal{G}}_{2n+1} = \{E_+^n, E_-^{n+1}\}$, $n \in \mathbb{Z}$, $\widehat{\mathcal{G}}_{2n} = \{H^n\}, n \in \{\mathbb{Z} - 0\}$.

7.2. Highest weight representations of $(sl_2)_q$

The highest weight vector v_0 of the fundamental representation of the quantum Lie algebra $(sl_2)_q$ [10] satisfies the relations $H_h v_0 = v_0$, $X_h^+ v_0 = 0$, $X_h^- v_0 = v_1$. Note that because of the definition of generators $(sl_2)_h$ (28) we have

$$[H, X_h^+]_h = [H, X_h^+] = 2X_h^+,\ [H, X_h^-]_h = [H, X_h^-] = -2X_h^-,\ [H, H_h]_h = [H, H_h] = 0.$$

The element H lies in Cartan subalgebra of $(sl_2)_q$. Therefore $H_h(Hv_0) = (Hv_0)$, $Hv_0 = \lambda v_0$. Using the definitions (28) we find that $H_h = \frac{2}{q+q^{-1}}\left(qX^+X^- - q^{-1}X^-X^+\right) = q^H(X_h^+X_h^- - X_h^-X_h^+)$, and the action of H_h on v_0 in such a form gives $q^H(X_h^+X_h^- - X_h^-X_h^+)v_0 = v_0$, and finally $X_h^\pm v_1 = q^{-\lambda}v_0$, where λ can be found from $\frac{2}{q+q^{-1}}\left(qX^+X^- - q^{-1}X^-X^+\right)v_0 = \frac{2}{q+q^{-1}}[H]v_0 = v_0$, with $\frac{2q(q^{\lambda}-q^{-\lambda})}{(q+q^{-1})(q-q^{-1})} = 1$. This procedure can be recurrently continued for all v_n in the basis of the representation. In section 7 we introduced a quantum affine algebra $(sl_2)_q^t$ as an affinization of $(sl_2)_q$. The highest weight vector of the i-th ($i = 1, 2$) fundamental representation of $(sl_2)_q^t$ possesses the properties similar to the properties of the highest weight vector of fundamental representation of $(sl_2)_q$. The action of $h_{1,2}$, $e_{1,2}$ and $f_{1,2}$ generators on highest weight vectors $v_0^{(1)}$ and $v_0^{(0)}$ is given by

$$\begin{gathered} h_1 v_0^{(1)} = v_0^{(1)},\ h_1 v_0^{(0)} = 0,\ h_0 v_0^{(1)} = 0,\ h_0 v_0^{(0)} = v_0^{(0)}, e_1 v_0^{(0)} = e_1 v_0^{(1)} = e_0 v_0^{(0)} = e_0 v_0^{(1)} = 0, \\ f_0 v_0^{(1)} = f_1 v_0^{(0)} = 0, f_0 v_0^{(0)} = v_1^{(0)}, f_1 v_0^{(0)} = v_1^{(0)}, \end{gathered} \tag{42}$$

where superscripts correspond to representation and subscripts are vector basis numbers. In the same way as for $(sl_2)_q$

$$e_1 v_1^{(1)} = q^{-H} v_0^{(1)},\ e_1 v_1^{(1)} = q^{-\lambda^{(1)}} v_0^{(1)},\ h v_0^{(1)} = \lambda^{(1)} v_0^{(1)},\ e_1 v_1^{(0)} = 0\ h = 1 \otimes H, \tag{43}$$

$$e_0 v_1^{(0)} = q^{-H} v_0^{(0)},\ e_0 v_1^{(0)} = q^{-\lambda^{(0)}} v_0^{(0)},\ h v_0^{(0)} = \lambda^{(0)} v_0^{(0)},\ e_0 v_1^{(1)} = 0, \tag{44}$$

$$H_1 v_1^{(1)} = (1 - 2qq^{-(\lambda^{(1)}-2)}) v_1^{(1)},\ H_1 v_1^{(0)} = 2q^{-1} q^{(2+\lambda_0^{(0)})} v_1^{(0)}. \tag{45}$$

8. Conclusions

In this paper we discussed affine Toda models formulated with the inclusion of higher then one grading subspaces for affine Lie algebras. The zero curevature condition applied to elements of connection involving higher gradign Lie algebra generators lead us to more suffisticated and interesting systems of equations describing dynamical systems that are closer (with respect to ordinary affine Toda models) to phenomena in nature. In particular, these systems of equations take into accound interractions of original Toda fields with matter fields. The group–theoretical way to treat higher grading affine Toda models allows us to find special solitonic solutions. In the quantum situation, corresponding methods using properties of quatnum groups and quantum Lie algebras [1] make possible constructions of special quantum soliton solutions.

9. References

[1] Delius G W and Hueffmann A 1996 J. Phys.A29
[2] Drinfeld V G 1985 Sov. Math. Dokl. 32
[3] Ferreira L A, Gervais J-L, Guillen, J S and Saveliev M V 1996 Nuclear Phys. B 470 1-2
[4] Jimbo M 1985 Lett. Math. Phys. 10, 63
[5] Kac V G 1990 *Infinite dimensional Lie algebras.* Third edition (Cambridge university Press, Cambridge)
[6] Leznov A N and Saveliev M V 1979 Lett. Math. Phys. 3
[7] Leznov A N and Saveliev M V 1992 *Group-Theoretical Methods for Integration of Non-Linear Dynamical Systems.* Progress in Physics Series, v. 15, (Birkhauser–Verlag, Basel)
[8] Mansfield P 1983 Nucl. Phys. B 222
[9] Olive D I Turok N and Underwood J W R 1993 Nucl. Phys. B401
[10] Zuevsky A. Quantum Lie algebra solitons. J. of Physics: Conf. Series (JPCS) 2014

IOP Conf. Series: Earth and Environmental Science **23** (2015) 012002 doi:10.1088/1755-1315/23/1/012002

The Behaviour of Laboratory Soil Electrical Resistivity Value under Basic Soil Properties Influences

Z A M Hazreek[1,7], M Aziman[2], A T S Azhar[3], W D Chitral[4,8], A Fauziah[5,9] and S Rosli[6]
[1,2,3] Faculty of Civil and Environmental Engineering, Universiti Tun Hussein Onn Malaysia, 86400 Batu Pahat Johor, MALAYSIA
[4] Research Center for Soft Soil, Universiti Tun Hussein Onn Malaysia, 86400 Batu Pahat Johor, MALAYSIA
[5] School of Civil Engineering, Universiti Sains Malaysia, 14300 Nibong Tebal Penang, MALAYSIA
[6,7] School of Physics, Universiti Sains Malaysia, 11800 USM Penang, MALAYSIA
[8] School of Architecture, Computing and Engineering, University of East London, Beckton E11 2RD London, ENGLAND
[9] Institute of Postgraduate Studies, Universiti Sains Malaysia, 11800 USM Penang, MALAYSIA

E-mail: hazreek@uthm.edu.my

Abstract. Electrical resistivity method (ERM) was a popular indirect geophysical tools adopted in engineering, environmental and archaeological studies. In the past, results of the electrical resistivity value (ERV) were always subjected to a long discussion and debate among the related parties such as an engineers, geophysicists and geologists due to its lack of clarification and evidences in quantitative point of view. Most of the results produced in the past was always been justified using qualitative ways which difficult to be accept by certain parties. In order to reduce the knowledge gap between those parties, this study has performed a laboratory experiment of soil box resistivity test which supported by an additional basic geotechnical test as referred to particle size distribution test (d), moisture content test (w), density test (ρ_{bulk}) and Atterberg limit test (LL, PL and PI). The test was performed to establish a series of electrical resistivity value with different quantity of water content for Clayey SILT and Silty SAND soil. It was found that the ERV of Silty SAND (600 – 7300 Ωm) was higher than Clayey SILT (13 – 7700 Ωm) due to the different quantity of basic soil properties value obtained from the basic geotechnical test. This study was successfully demonstrated that the fluctuation of ERV has greatly influenced by the variations of the soil physical properties (d, w, ρ_{bulk}, LL, PL and PI). Hence, the confidence level of ERV interpretation will be increasingly meaningful since it able to be proved by others parameter generated by laboratory direct test.

1. Introduction

Geotechnical site investigation (SI) was carried out to determine the properties of the geomaterials (soil and rock) which importantly used in design and construction stage. Generally, it involves two stage involving surface and subsurface exploration. Physical mapping (e.g. geological mapping) was always conducted during the surface exploration while soil sampling and laboratory test was usually performed during the subsurface exploration. Several techniques were commonly used in the SI involving conventional and alternative approach. Drilling technique was a classical SI method which obtained the SI data in good accuracy due to its nature of direct test. However in certain cases (e.g. large scale project), most of the conventional methods will faced several constraints which related to the increasing of project cost and time. Furthermore, the information obtained was a single point data and the interpolation between a large boreholes spacing can lead to increase the degree of uncertainties of the subsurface profile investigated [1] – [3]. As a result, the solutions to these challenges will require multidisciplinary research across the social and physical sciences and engineering [4]. Hence, geophysical method offers the chance to overcome some of the problems inherent in more conventional ground investigation techniques [5].

Today, geophysical method such as electrical resistivity, seismic, gravity, ground penetration radar, magnetic, etc has improve lots due to the continuous rapid development of electronics technology thus increasingly adopted as an alternative tool in SI. Geophysical methods can be implemented more quickly and less expensively and can cover larger areas more thoroughly [6] – [9]. The electrical resistivity method was among the most popular geophysical tools applied in geotechnical site investigation for subsurface mapping tools. This method was used to detect the depth and thickness of soil and rock, bedrock and overburden materials [10], groundwater [11] and [12], cavity and boulder [13], etc. Most of the project involving ERM was performed by geophysicists, engineers and geologists who have their own strength and weakness based on their own background areas. For example, geophysicist was a truly championed in geophysical field which allow them to performed the electrical resistivity data acquisition, processing and interpretation very good while the civil engineers has not being fully explored about this method due to their lack of exposure and expertise in this area.

As reported by [1], some geophysical results and conclusions are difficult to deliver in sound and definitive ways due to some of geophysicists who attempt to hide their expertise for business reasons. Conventional geophysical interpretation such as ERV is too much dependent subjectively on anomaly and past geomaterials reference charts and tables thus possibly considered weak in term of its reliability due to the qualitative point of view approaches. For example, the properties of ERV can be found to be dissimilar even for the same particular type of geomaterial in many past references. Moreover, conventional reference tables of geomaterials used for anomaly interpretation sometimes was difficult to decipher due to its wide range of variation and overlapping values [14]. As a result, a strong verification is vital to explained its behaviour which otherwise have been traditionally conclude based on a qualitative approach depending on the experience of the expert [15]. This scenario has proved that the geophysical properties ambiguity are still need to be discover from several perspectives. According to [5], geophysicists still possess little appreciation from the engineer's point of view and lack the knowledge of the science in soil mechanics.

Hence, this study was performed to established a series of laboratory soil electrical resistivity properties and basic physical properties (BPP) under controlled environment with the aim to reduce some black box and ambiguities between all related parties via relationship of ERV and basic physical properties of soil with particular reference to moisture content, density and soil grain size characteristics. Laboratory test used in this study was able to reduced soil uncertainties thus able to produced significant results of the soil ERV behaviour due to its control environment (temperature, humidity and properties known). Furthermore, this fundamental study can possibly increased the confidence level of engineers regarding the ERV interpretation which importantly use as an initial study during the subsurface profile characterization.

2. Methodology

All tests were performed on two different types of soil (fine and coarse soil) using laboratory based geophysical and geotechnical experiments. Soil classification tests such as dry and wet sieve, Atterberg limit, moisture content and density based on [16] was carried out. Electrical soil box resistivity test was performed using Nilsson model 400 soil resistance meters and Miller's soil box apparatus by mixing an original mass of 1500 g of oven dried soil with 1-3 % of distilled water and tested repeatedly at 25 times (each test used 15, 30 and 45 ml of distilled water based on percentage of water used for 1500 g of original oven dried soil). For example, a 15 ml of distilled water was added consistently and mixed thoroughly into the originally loose oven dried soil and continuously tested with each increment of distilled water added using soil box resistivity meter for 25 determinations of ERV. After that, the same procedure was repeated using a 30 and 45 ml of distilled water which applied to the new 1500 g of originally loose oven dried soil. The test was continuously repeated at 25 times due to the consistency of ERV (ERV has demonstrate insignificant changing when it reach 25 numbers of test). Miller's soil box apparatus consisted of a 4 pin arrangement consisting of 2 current pin (both being located at the end of the soil box) and 2 potential pin (these being located along the middle of the soil box). The function of two end current pins were to inject direct current (DC current) into the soil while the other two potential pins were used to measure a potential difference for calculating the soil resistivity value. Soil moisture content and density was taken immediately after the soil resistivity was measured. Moisture content test was determined for two samples from each soil box test for final averaging purposes. All results obtained from the experiments were also analyzed using a statistical regression method. Overall methodology of this study was given in the Figure 1. As referred to in [16], the following equations (1-3) were used to calculate the resistivity value, bulk density and moisture content.

$$\rho = RA/L \tag{1}$$

where A is the cross-sectional area of the sample, L is the length of the sample between the electrodes and R is the mean resistance of the soil sample ($R=V/I$)

$$\rho_{bulk} = m/V \tag{2}$$

where m is the mass of the soil specimen (solids + water) and V is the volume of the test specimen (total volume)

$$w = ((m_2 - m_3)/(m_3 - m_1)) \text{ x } 100 \tag{3}$$

where m_1 is the mass of container, m_2 is the mass of container and wet soil and m_3 is the mass of container and dry soil

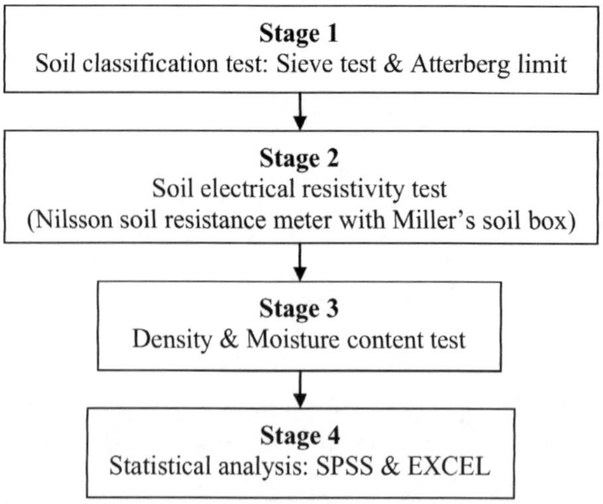

Figure 1. Analysis path to infer the behaviour of laboratory soil electrical resistivity value under basic soil properties influences.

3. Results and Discussions

All results presented and discussed are based on laboratory electrical resistivity value (ERV), basic physical properties of soil and relationship of field ERV with moisture content (w), density (ρ) and grain size of soil (d). All results are presented in Table 1 and Figure 2 – 4.

3.1. *Electrical resistivity value (ERV)*

ERV was determined by measuring the potential difference at points on the ground surface which caused the propagation of direct current through the subsurface [17]. The ERV obtained in Table 1 was originally established from the laboratory soil box test using two different types of soil under loose condition. It was found that the ERV of Silty SAND (600 – 7300 Ωm) has demonstrate a higher ERV compared to the Clayey SILT (13 – 7700 Ωm) due to the different composition of basic physical properties of soil such as solid, water and air. Electrical propagation in soil is largely electrolytic process by flowing in connected pore spaces and along grain boundaries of geomaterial [18].

3.2. *Grain size fraction, moisture content and density*

This study has performed using fine and granular soil in order to demonstrate the behaviour of the ERV under different type of soil which contained different composition of physical and chemical properties. Based on sieve and Atterberg limit test, it was found that the fine soil sample was classified as Clayey SILT (mixture of both granular and fine particle with domination of fine particle) while granular soil sample was known as Silty SAND (mixture of both granular and fine particle with domination of granular particle) based on engineering soil classification. All sieve analysis results of soil specimen tested from both soils has shown some great variation in terms of grain size quantification due to the natural heterogeneity features of soil. Detailed results obtain in Table 1 was originally extracted from particle size distribution curve (PSD) presented in Figure 2.

It was found that moisture content (w) value of Clayey SILT (7.76 – 100 %) has a higher range compared to the Silty SAND (17.66 – 61.07 %). For density results, it was found that Silty SAND (ρ_{bulk} = 1.77 – 2.66 Mg/m^3) has a higher bulk density value compared to the Clayey SILT (ρ_{bulk} = 1.00 – 1.69 Mg/m^3). In soil mechanics and geotechnical engineering, soil density was basically described using bulk density (ρ_{bulk}) and dry density (ρ_{dry}). Bulk density was defined by total mass of solids and water per total volume while dry density was defined by mass of solids per total volume. Quantities of densities provide a measure of the material quantity related to the space amount it occupies [19]. Based on Figure 3 and 4, it can be observed that the densities of each soil were relative to the moisture content variations by showing general relationship of densities was linearly proportional with the moisture content. Higher quantity of water will increased the total soil weight thus increasing its density and vice versa. Quantity of fine particles from Clayey SILT was greater than coarse particle which allow more water to be absorb more in order to increase its densities. As for Silty SAND, the relationship was strongly believed to be influenced by the domination of water and coarse grain particle. Coarse grain material has a lower ability to absorb water but higher in term of its mass. Hence, the value of moisture content (w) of Silty SAND was lower than Clayey SILT but higher in term of bulk density which caused by a higher mass of coarse particle and also the present of water. Generally, both types of soils have demonstrated different variation of moisture content and density due to the dissimilarity composition of grain sizes fraction.

3.3. *Behaviour of electrical resistivity value due to the basic physical properties of soil*

This study has demonstrated the relationship of Clayey SILT and Silty SAND due to the grain size fraction, moisture content, density and variations. The behaviour of the ERV due to the BPP were also being analyzed and presented using statistical regression method in Figure 3 and 4.

Based on Table 1, it was found that the ERV of Silty SAND (600 – 7300 Ωm) was higher than Clayey SILT (13 - 7700 Ωm). Based on grain size fraction, Silty SAND was dominantly composed from coarse soil (60.09 %) while Clayey SILT was mainly composed from fine soil (60.09 %). According to previous research, the ease of current to propagate within each geomaterial can be influenced by grain size fraction as fine particle will allowed current to easily propagate thus producing low resistivity value while coarse grain fraction will produced high resistivity value due to the difficulty of current propagation. According to [19], current may propagate in geomaterials based on electrolysis conduction where the propagation is by ionic conduction from amount and conductivity of dissolved minerals present. In mineral composition of soil, Clayey SILT has a higher composition of fine soil with particular reference to clay minerals such as kaolinite, illite, montmorillonite and vermiculite which allowed the ease of the current propagation in the soil tested thus producing low resistivity value compared to the Silty SAND which has low composition of clay with high composition of coarse soil. Furthermore as reported by [20], resistivity data exhibits a low value for a fine soil such as clayey and silty while the coarser soil such as sand and gravel will produce a higher resistivity value.

Based on this study, soil moisture content and density has demonstrated great influence to the behaviour of ERV. It was found that both correlations (ERV with w and ρ_{bulk}) showed a curvilinear trend as given in Figure 3 and 4. According to Figures 3 and 4, moisture content and bulk density value was higher with the decrease of soil electrical resistivity value due to the presence of more water content in soil tested. This finding confirmed the past findings which indicated the moisture content value will increase with the decreasing of soil resistivity value [21] – [25] which can be represented using general relationship of ERV ∞ $1/w$. When water was added continuously from dry soil until it reach saturated and oversaturated condition, the ease of current propagation in soil will be increased, thus producing a high to low resistivity value. As reported by [24], ions in pore fluid finds hard to propagate in the low moisture content soil thus producing a low soil conductivity which will cause an increasing of resistivity value. The ability of Clayey SILT to absorbed more water (16 – 100 %) due to its high quantity of fine soil (Clay and Silt) has greatly influenced to dropped the ERV value compared to the Silty SAND which has less fine soil (lower water absorbance (17.66 – 61.07 %) due to less fine soil with high coarse soil). Hence, the ERV of Clayey SILT can be obtained lower than Silty SAND due to its higher water absorbance ability which related to the composition of the soil particle.

Both soils were tested under loose condition which influenced the ERV due to its varying soil density. Generally, both soils were demonstrated that ERV was high at low soil density and vice versa. This condition has shown that under low soil density, soil porosity was filled by air thus producing high ERV. According to [13], air filled void posses a higher resistivity value compared with the water filled void. Both soils were tested from dry–moist–saturated–oversaturated condition using 1-3 % of water continuously until 25 times under loose condition. The quantity of water was greatly increased thus increasing the soil bulk density which finally produced low ERV when the experiment has passed the dry-moist state at saturated-oversaturated condition. Clayey SILT composed of higher fine grain particles was able to absorb more water compared to the Silty SAND. Hence, this phenomenon was possibly has affected the relationship between ERV and densities. Commonly, it was expected that the ERV was supposedly to be high due to the higher soil densities. However in Clayey SILT, this hypothesis was unsuccessfully due to the presence of more water within the fine soils with particular reference to clay and silt particles. Hence in Clayey SILT case, higher density was associated with a higher moisture content thus producing a low ERV which can be represent by ERV ∞ $1/\rho$. In other words, the higher moisture content causes easily a flow of current within the soil which finally produced a lower ERV. However the value for soil density of Clayey SILT was lower than Silty SAND due to its higher quantity of fine soil. A higher ERV will be produced due to the higher value of soil densities which associated by higher quantity of granular soil with particular reference to gravel particles demonstrated by Silty SAND. As reported by [26], the bulk resistivity of soil will increase with the grain size increment since it offers more resistance to the ionic current flow. Moreover, higher granular soil will produced lower moisture content which also contributes to the increasing of ERV. Hence, general relationship between ERV and soil densities of Gravelly SAND can be found as ERV ∞ ρ.

This study found that the ERV will continuously decrease as water was continuously added. Despite an overall continuous decrease in resistivity value, a small inconsistency of value reduction from high to low was also recorded. This inconsistency ERV was caused by small inconsistent quantity of soil tested for each percentage of water added due to the difficulty of soil handling from dry-moist-saturated-oversaturated. The experiment was carefully performed by filling the soil (mixed thoroughly with water) inside the box without being compacted (loose condition) with the lowest possible void left. However, the workability of soil inside the box was difficult especially when it was at moist to saturated state where the soil was in a highly cohesive condition. The problems continued when it reached saturated to oversaturated state where the quantity of soil tested can be varies due to the large quantity of water added. Furthermore in some cases, those relationships presented will turn inversely especially when the properties obtained were almost similar to each other. Hence, other major non similar properties will take placed to influence the field ERV. Based on [24], detailed study related to the field condition such as porosity, degree of saturation, salt concentration in pore fluid, grain size, size gradation, temperature and activity can produce more accurate correlation performed from the laboratory experiment.

Resistivity value was highly influenced by pore fluid and grain matrix of geomaterials [27]. As reported by [28], a soil's ERV generally varies inversely proportional to the water content and dissolved ion concentration as clayey soil exhibit high dissolved ion concentration, wet clayey soils have lowest resistivity of all soil materials while coarse, dry sand and gravel deposits and massive bedded and hard bedrocks have the highest ERV. Furthermore a decrease of ERV was results from an increased of metal ions or inorganic elements in geomaterials [29]. According to [30], soil parameters determined in grain size analysis could replicate the variety of resistivities obtained on the site very well. Hence, this study was able to explained varying values due to the variation of selected soil physical state. In other words, BPP can strongly influence the field ERV due to soil composition variation such as relative to the quantity of solid, air and water.

Table 1. ERV and basic physical properties of soil.

Soil Type	Clayey SILT				Silty SAND			
Electrical resistivity value, ρ (Ωm)	13 - 7700				600 - 7300			
Moisture content, w (%)	7.76 - 100				17.66 – 61.07			
Bulk density, ρ_{bulk} (Mg/m^3)	1.00 – 1.69				1.77 – 2.66			
Particle size analysis, d (μm – mm, %)	Clay	Silt	Sand	Gravel	Clay	Silt	Sand	Gravel
	24.82	54.08	16.51	4.59	3.40	36.51	47.35	12.74
	78.90		21.10		39.91		60.09	
Atterberg limit (%)	LL	PL	PI		LL	PL	PI	
	51.90	39.80	12.10		38.40	31.15	7.25	

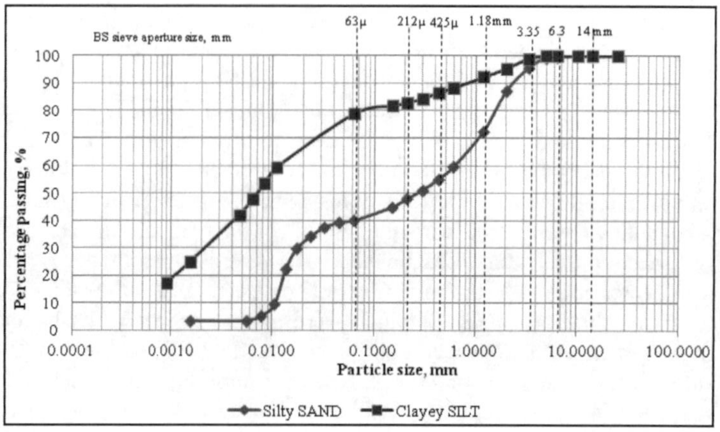

Figure 2. PSD curve for Clayey SILT and Silty SAND from dry and wet sieve analysis.

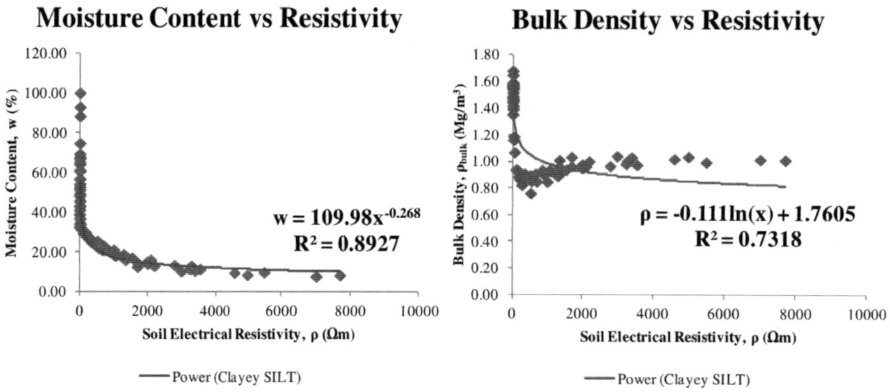

Figure 3. Moisture content (left) and density (right) correlation with ERV for Clayey SILT.

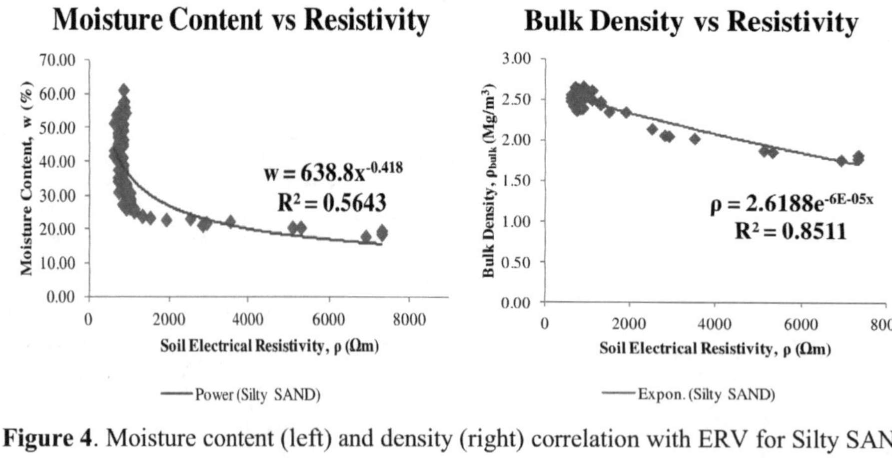

Figure 4. Moisture content (left) and density (right) correlation with ERV for Silty SAND.

Conclusion

The ERV from laboratory test was successfully established on Clayey SILT and Silty SAND soil. The ERV was largely influenced by the variations of basic physical properties of soil. This study has reduced few of the black boxes (uncertainties) of the ERV interpretation ambiguity based on BPP influences with particular reference to grain size fraction, moisture content and density of soil. This study has demonstrated that the integration between ERV with BPP can contribute a meaningful contribution especially during the interpretation stage since it was able to explained and justify the behaviour of ERV based on BPP quantitative approach.

Acknowledgments

This work was funded by Universiti Tun Hussein Onn Malaysia (ERGS from vote 40) and Universiti Sains Malaysia. First author wish to acknowledge gratefully to supervisors and research members for their tremendous guidance, work and cooperation.

References

[1] Abidin M H Z, Ahmad F, Wijeyesekera D C, Saad R and Baharuddin M F T 2013 Soil Resistivity Measurements to Predict Moisture Content and Density in Loose and Dense Soil *Applied Mechanics and Materials.* **353-356** 911-917

[2] Godio A, Strobbia C and De Bacco G 2006 Geophysical Characterisation of a Rockslide in an Alpine Region, *Engineering Geology.* **83** 273-286

[3] Mauritsch H J, Seiberl W, Arndt R, Römer A, Schneiderbauer K and Sendlhofer G P 2000 Geophysical Investigations of Large Landslides in the Carnic Region of Southern Austria, *Engineering Geology.* **56** 373-388

[4] Fragaszy R, Santamarina J, Amekudzi A, Assimaki D, Bachus R, Burns S, Cha M, Cho G, Cortes D, Dai S, Espinoza D, Garrow L, Huang H, Jang J, Jung J, Kim S, Kurtis K, Lee C, Pasten C, Phadnis H, Rix G, Shin H, Torres M and Tsouris C 2011 Sustainable development and energy geotechnology — Potential roles for geotechnical engineering *KSCE Journal of Civil Engineering.* **15** 611-621

[5] Clayton C R I, Matthews M C and Simons N E 1995 *Site Investigation* (UK: Blackwell Science Ltd)

[6] Khatri R, Shrivastava V K and Chandak R 2011 Correlation between vertical electric sounding and conventional methods of geotechnical site investigation *Int. Journal of Advanced Engineering Sciences and Technologies.* **4** 042-053

[7] Liu C and Evett J B 2008 *Soils and Foundation* (New Jersey: Pearson International)

[8] Godio A, Strobbia C and De Bacco G 2005 Geophysical characterisation of a rockslide in an alpine region *Engineering Geology.* **83** 273-286

[9] Cosenza P, Marmet E, Rejiba F, Jun Cui Y, Tabbagh A and Charlery Y 2006 Correlations between geotechnical and electrical data: A case study at Garchy in France *Journal of Applied Geophysics.* **60** 165-178

[10] Saad R, Muztaza N M and Mohamad E T 2011 The 2D Electrical Resistivity Tomography (ERT) Study for Civil and Geotechnical Engineering Purposes *Electronic Journal of Geotechnical Engineering.* **16** 1537-1545

[11] Hamzah U, Samsudin A.R and Malim E P 2007 Groundwater investigation in Kuala Selangor using vertical electrical sounding (VES) surveys *Environmental Geology.* **51** 1349-1359

[12] Hamzah U, Yaacup R, Samsudin A. R and Ayub M S 2006 Electrical imaging of the Groundwater Aquifer at Banting, Selangor, Malaysia *Environmental Geology.* **49** 1156-1162

[13] Jusoh Z 2010 Application of 2-D resistivity imaging and seismic refraction technique in subsurface investigation for civil engineering, *M.S. thesis*, Sch. Physics, Science Univ., Penang, Malaysia

[14] Solberg I L, Hansen L, Rønning J S, Haugen E D, Dalsegg E and Tønnesen J 2011 Combined Geophysical and Geotechnical Approach to Ground Investigations and Hazard Zonation of a

Quick Clay Area, Mid Norway *Bulletin of Engineering Geology and the Environment.* **71** 119-133

[15] Abidin M H Z, Wijeyesekera D C, Saad R and Ahmad F 2013 The Influence of Soil Moisture Content and Grain Size Characteristics on its Field Electrical Resistivity *Electronic Journal of Geotechnical Engineering.* **18/D** 699-705

[16] British Standard 1377 1990 *Methods of test for Soils for Civil Engineering Purposes*

[17] Burger H R, Sheehan A F and Jones C H 2006 *Introduction to Applied Geophysics* (New York: W.W. Norton & Company)

[18] Anita T, 2005 Water content and porosity estimated from ground-penetrating radar and resistivity *Journal of Applied Geophysics.* **58** 99-111

[19] Telford W M, Geldart L P and Sheriff R E 1990 *Applied Geophysics* (Cambridge: Cambridge University Press)

[20] Abidin M H Z, Saad R, Ahmad F, Wijeyesekera D C and Baharuddin M F T 2012 Integral Analysis of Geoelectrical (Resistivity) and Geotechnical (SPT) Data in Slope Stability Assessment *Academic Journal of Science.* **1** 305-316

[21] Siddiqui F I and Osman S B A S 2012 Integrating Geo-Electrical and Geotechnical Data for Soil Characterization *International Journal of Applied Physics and Mathematics.* **2** 104-106

[22] Ozcep F, Yildirim E, Tezel O, Asci M and Karabulut S 2010 Correlation between electrical resistivity and soil-water content based artificial intelligent techniques *International Journal of Physical Sciences.* **5(1)** 047-056

[23] Cosenza P, Marmet E, Rejiba F, Cui Y J, Tabbagh A and Charlery Y 2006 Correlations between geotechnical and electrical data: A case study at Garchy in France *Journal of Applied Geophysics.* **60** 165-178

[24] Rinaldi V A and Cuestas G 2002 Ohmic Conductivity of a Compacted Silty Clay *Journal of Geotechnical and Geoenvironmental Engineering.* **128(10)** 824-835

[25] Tezel O and Ozcep F 2003 Relationships of electrical resistivity and geotechnical parameters," *Proc. Conf. on Earth Sciences and Electronics (Istanbul, Turkey, 2003)*

[26] Sudha, K., Israil, M., Mittal, S. and Rai, J., Soil characterization using electrical resistivity tomography and geotechnical investigations. *Journal of Applied Geophysics*, Vol. 67, 2009, pp. 74-79.

[27] Griffiths D H and King R F 1981 *Applied Geophysics for Geologist and Engineers-The Element of Geophysical Prospecting* (Oxford: Pergamon Press)

[28] Liu C and Evett J B 2008 *Soils and Foundation* (New Jersey: Pearson International)

[29] Jung Y, Lee Y and Ha H 2000 Application of Electrical Resistivity Imaging Techniques to Civil and Environmental Problems *Use of Geophysical Methods in Construction*

[30] Friedel S, Thielen A and Springman S M 2006 Investigation of a slope endangered by rainfall-induced landslides using 3D resistivity tomography and geotechnical testing *Journal of Applied Geophysics.* **60** 100-114

Structural Balancing vs Horizon Flattening on Seismic Data: Example from Extensional Tectonic Setting

S.N. Fathiyah Jamaludin[1], A.H. Abdul Latiff [2],D.P. Ghosh[3]

[1,2,3]Center for Seismic Imaging,
Universiti Teknologi PETRONAS,
Bandar Seri Iskandar, Ipoh,
31750 Malaysia

E-mail: fathiyah.jamaludin@yahoo.com[1]

Abstract: Two techniques which are structural balancing and horizon flattening have been applied in this work to verify and validate the seismic interpretation performed on a few extracted 2D seismic cross-sections. Both techniques are useful to provide a mean of understanding what had happened or at least to imagine what had happened during the time of deposition for each of the interpreted horizons. Structural balancing technique is somehow more difficult, time-consuming and requires other information such as rocks' densities before it can be practiced. Horizon flattening, on the other hand is a straight-forward technique available in most interpretation software. Comparing the results from both techniques had shown us that structural balancing, despite its difficulties and time-consuming had provided a more geologically sound interpretation. It allows the interpreters to understand and imagine the relationship of faulting with sedimentation at a particular time. However, horizon flattening only provides a general overview on the morphology and environment of deposition for the flattened horizons. Despite all, both techniques require a familiarity with the interpretation and structural software and what is important is an interpreter who can make his or her interpretation geologically acceptable and logic.

Keywords: Structural Balancing, Horizon Flattening, Seismic Interpretation, Structural Reconstruction, Seismic

1. Introduction

Interpreting a seismic cross-section may vary differently from each of us. Geologists, geophysicists, petroleum engineers and others are potential users of seismic data and they may interpret seismic data according to their objectives of the projects. It is a common practice for a geoscientist to conduct stratigraphic and structural interpretation on the seismic data, be it a 2D or 3D seismic dataset. Seismic data is not 100% convincing if it is to stand alone. Thus, availability of wells data and core plugs or core logs will be very beneficial to validate interpretation performed on the seismic data. However, petroleum-industry seismic data that are commonly donated to research groups in universities or research centers do not necessarily come along with wells and core data. For the

[1] Corresponding author.

purpose of interpreting the stratigraphic and structural context based on only seismic data, application of structural balancing and horizon flattening will provide a mean of justification for the interpretation done. This paper is aim to compare the advantages and disadvantages gathered from structural balancing and horizon flattening techniques. We have applied these two techniques to rectify the seismic interpretation done on the data from extensional tectonic setting in the Central Luconia Province, offshore Sarawak, Malaysia.

2. Methodology

Seismic interpretation has been performed on few extracted seismic cross-sections. Cross-section balancing and horizon flattening are conducted on the interpreted cross-sections to validate the interpretation, as well as reveals the features at the particular time of depositions. Below are the detailed descriptions for the methods practiced in this research.

A. Seismic Interpretation

Interpretation of seismic data began with creation of synthetic seismogram, as an aid to compliment geological information from well data in depth with the geophysical information from the seismic data in time. Check shot and sonic logs from two wells were used to provide time to depth relationship. Standard procedures for synthetic seismogram creation have been followed. Three statistical wavelets were tested, with 30Hz, 40Hz and 45Hz dominant frequency, zero phasing and standard SEG polarity conversion. Statistical wavelet of 40Hz give the best result for the creation of synthetic seismogram, thus was used in this research. Seismic interpretation including stratigraphic and structural interpretation was performed based on few seismic attributes. These include True Amplitude, Relative Acoustic Impedance, Coherence-Variance and Curvature attributes. Seismic reflection geometries were also applied during interpretation. Seismic reflection geometries were recognized and interpreted for the purpose of understanding the environment of deposition for each interpreted sequences.

B. Structural/Cross-section Balancing/Restorations

Balancing a section is about putting back the point where deformation initiates to its undeformed state. It is an aid to validate our interpretation, be it seismic interpretation or interpretation based on outcrop by applying certain geometric rules. It involves removing the effects of fault displacements, folding associated with faulting and flexural slip and volume loss generated from compaction and erosion. A balanced cross section does not necessarily provide a correct restoration (Fossen [1]), because there could have been layer thickening involve and also growth of carbonate strata on the layers. In this work, cross-section or structural balancing was performed using MOVE software. MOVE is structural modeling and analysis software, made available by Midland Valley Exploration. Briefly, there are seven computer algorithms used in Kinematic Modeling for balancing a cross-section, depending on the tectonic settings. In the case of extensional tectonic setting in offshore Sarawak, two balancing techniques are applied, Simple Shear and Rigid Block Restorations. Simple Shear and Rigid Block Restorations techniques are most suitable for the purpose of balancing cross-sections in extensional setting as it involves growth listric faults, dominos fault system and non-planar normal fault.

1) Simple Shear

The Simple Shear algorithm models the relationship between fault geometry and hanging wall deformational features. It can be divided into Vertical Simple Shear (VSS) and Oblique Simple Shear (OSS) techniques. Using VSS, the shear plane is assumed to make an angle of 90° to the regional horizontal plane, while OSS on the other hand is simple shear along the horizontal planes at angles other than 90°. Based on the assumptions that the last process of hanging wall collapse, the last fault thought to have developed is restored first (Bland, et al. [2]).

2) Rigid Body Restoration

Apart from simple shear technique, in situation where displacements are too small or involved tilted blocks, rigid- body restoration technique has been adapted to the cross-section restoration. Rigid-body restoration preserved the original lengths and angles within the blocks of deformed cross-section. This is usually applied to the simplest case of cross-section restoration where fault blocks behave as rigid blocks during deformation, so that rigid rotation and translation are involved in the same restoration step. The domino fault system where both translation (displacement/offset) and rotation (anticlockwise) are involved is the best example for restoration using this method. Each faulted block can be restored either by firstly rotating the block so they become horizontal, before removing the offsets; or secondly by removing the offsets created by the fault, followed by rotation. The exact ways of deformation to take place in the case of rotated fault-blocks were not to worry, as both translation and rotation may actually formed simultaneously (Fossen [1]) in the real process. In rigid body restoration process, if the blocks do not fit together, it is preferable to leave gaps between them, rather than overlapping the blocks (Groshong [3]).

C. Horizon Flattening

One of the available simple and fast techniques to help geologist and geophysicist making decisions during interpretation is horizon flattening. Horizon flattening is the digital version of the interpreter taking a folded paper section and overlaying one part on another to check the character and correlation (Bland, et al. [2]). Horizon flattening technique is available in most geological interpretation software. It provides a good tool to predict the continuity of the stratigraphic and structural elements in the underneath layers.

3.0 Results & Discussions

Twelve horizons, including sea bottom with 5 seismic sequences were interpreted prior to seismic cross-section restoration. The sections were restored in three stages which include Recent (sea bottom) - Early Pliocene sediment removal, Late- Middle Miocene Restoration and Early Miocene/ Late Oligocene Restoration. Each restoration stages include few intermediate restoration phases to compensate small displacements created by small-scale faults within the seismic stratigraphic units. The time interval for the restoration involved is from sea bottom (0 Ma) to Late Oligocene (34 Ma).

Figure 1 below shows the results of seismic interpretation over two Miocene carbonate platform in offshore Sarawak. However, for structural balancing and horizon flattening, Platform EX (Figure 2) was excluded from this technique, due to its' small size of line distribution. Structural balancing and horizon flattening were performed on Platform FY (Figure 1) only. There are three major restoration steps, which include Step 1: Recent- Early Pliocene, Step 2: Late-Middle Miocene and Step 3: Early Miocene to Late Oligocene. Each restorations step involved intermediate balancing to compensate the small scale branching faults within each sequences.

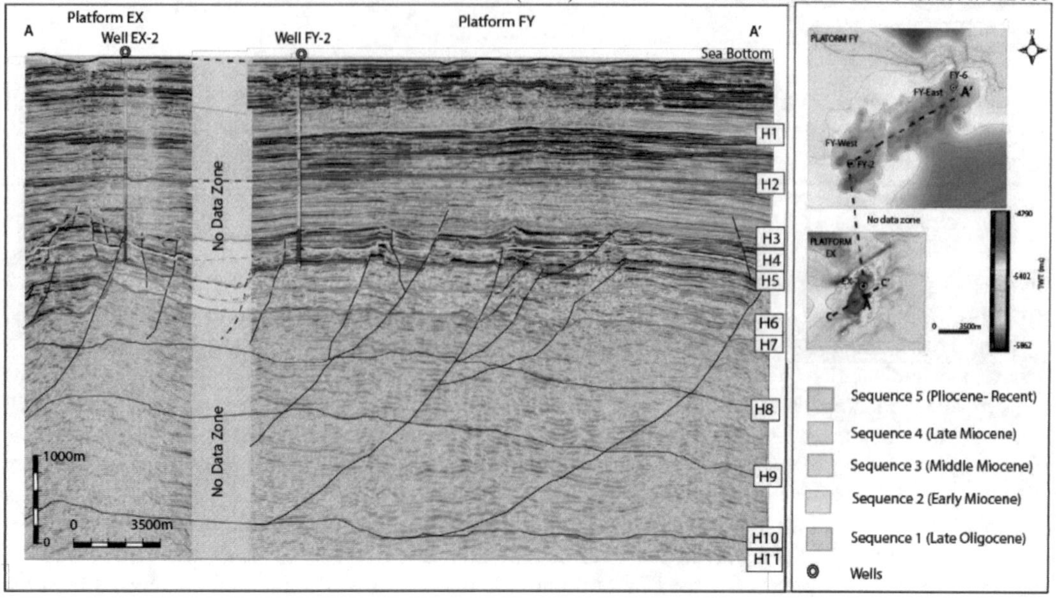

Figure 1: Seismic interpretation based on structural and stratigraphy characteristics from Line A-A' crossing Platform FY and EX. Interpreted horizons are marked as H1 to H11. Map is showing the location of the line.

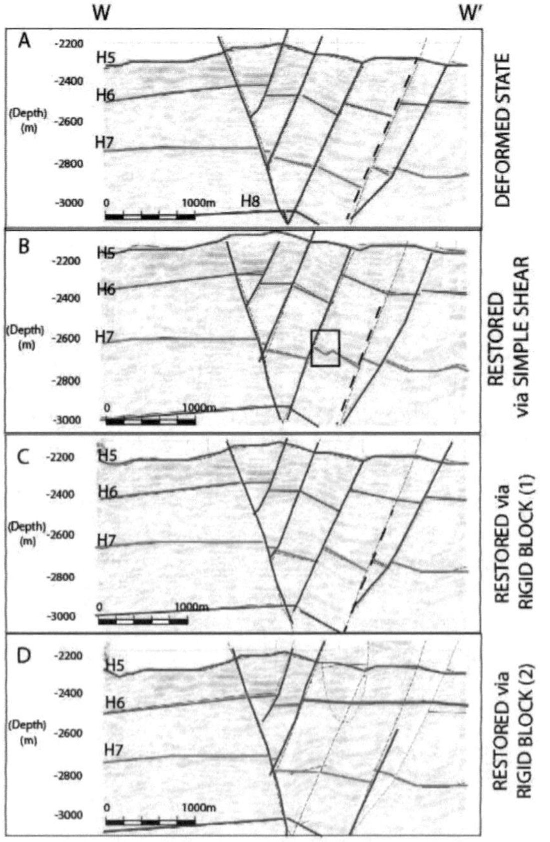

Figure 2: Small portion of Inline WW', showing antithetic branching faults within Horizon 5-7. Two restorations techniques were tested. The restored fault is corresponding to the dash line. Restoration via simple shear produced an artificial bending (in square) at H7 while restoration via Rigid-Body technique matches H6 in the hanging wall and foot wall. Continuation of restoring in D using Rigid-Body technique had fixed the offsets and flattens H6 and H7.

Horizon flattening was performed only on Horizon 2 to Horizon 6. The deeper horizons (H7-H11) do not serve a significant seismic resolution to allow horizon flattening. During Step 3 restoration, all three balancing techniques (VSS, OSS and Rigid Body) for extensional setting have been applied. Step 3 restorations involved balancing few sets of antithetic faults within domino fault systems. Balancing an antithetic fault pose a special dilemma (Rowan and Kligfield [4]), because it terminates at another fault without offsetting it, and displacement along it will therefore decrease to zero, thus none of the standard restoration techniques would resolved this.

Figure 2 shows a small portion of antithetic fault in Inline WW' in depth section. The displacements of these antithetic faults were removed using rigid block restoration and vertical simple shear techniques. Both restoration techniques produce different results. With vertical simple shear, artificial bending occurs above the fault termination. The most likely explanation for this situation is that complex deformation in hanging walls of the antithetic faults is inadequately modelled by the standard restoration algorithms, therefore, must be carried out with more general area-conservation methods. To simply restore the antithetic faults, rigid block restoration method is more appropriate in this case.

Horizon flattening had revealed significant features present at a particular time. However, this process has a number of drawbacks that always lead to overlooking in the particulars such as distortions in the image and artefacts of the flattening process (Bland, et al. [2]). Artefacts and distortions from flattening processes can significantly mislead interpreter if not recognized. Structural balancing technique on the contrary provides a validation in the interpretation applied to particular cross-sections. Variables techniques through structural restoration such as sediments removal via decompaction technique and isostatic adjustment on the faulting via move-on-fault technique have a geologically valid way of looking on the geological history, with reference to present day acquired seismic image.

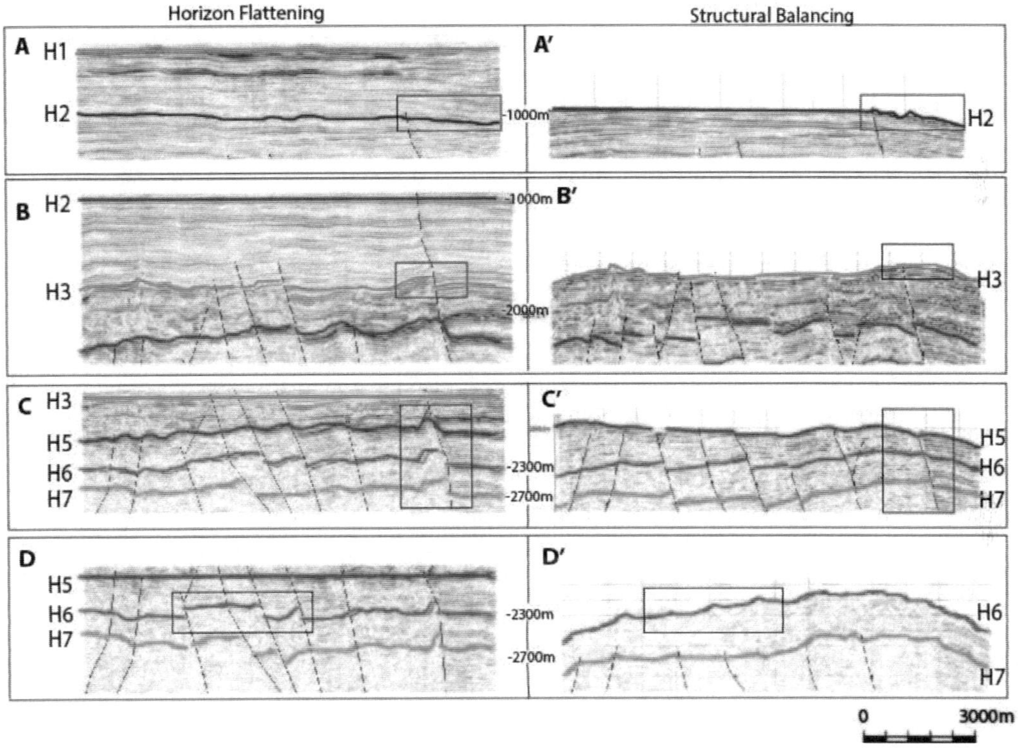

Figure 3: A: Flatten Horizon 1 (H1) to see the morphology of underneath Shale Layer bounded by Horizon 2. A': Exposed morphology of Shale Layer in Horizon 2 (H2) with undulating effects on the hanging wall (small box). B: Flatten Horizon 2 (H2) exposed the morphology of Horizon 3 (H3) with still obvious displacement on the foot wall and hanging wall. B': Restored Horizon 3 (H3) exposed hump-back features (in small box) which are not seen on Horizon 3 in Figure 3B. C: Flatten Horizon 3 (H3) to see the morphology of deeper horizons. Artefacts (in small box) developed on the foot wall of Horizon 5-7 resulted from flatten Horizon 3. C': A balanced section of Horizon 5 (H5) shows no paleo-topography on the foot wall as per seen in the flattening horizon in Figure 3C. D: Flatten Horizon 5 (H5) give a dragging effects on Horizon 6 (H6) (in small box) and create larger displacement. However, this is not seen in the balanced section of Horizon 6 (H6) in Figure 3D'.

In order for us to see the features present at a particular time, the top (younger) horizon will be flatten. Horizon 1 (H1) has been flatten to reveal the situation during deposition of older layers in Horizon 2 (H2). Meanwhile, structural balancing involved decompaction and sediment removal from sea bottom to Horizon 1 (H1), followed by restoration of foot wall block in Horizon 2 (H2) had brought us back to the time when Horizon 2 (H2) deposited. Referring to Figure 3A and 3A', features in the small rectangular has highlighted that result from structural balancing had formed an undulating effects on Horizon 2, which might be an artefact. In this case, horizon flattening brings out a more logical result compared to structural balancing technique.

Flattened Horizon 2 in Figure 3B has exposed depositional features during deposition of Horizon 3. Displacements on the foot wall and hanging wall are seen during deposition of Horizon 3. However, through structural balancing technique, offsets seen on Horizon 3 have been fixed thus revealing a hump-back structure as highlighted in the rectangular. Structural restoration on Horizon 2 produced a more valid and geologically accepted result compared to horizon flattening technique.

Comparison of these two techniques on Horizon 5, 6 and 7 is shown on Figure 3C and 3C'. Flattening of Horizon 3 had created mounded features on the foot wall of these horizons. Initially, we thought it is stack of reefal group growth. To validate our interpretation, we applied structural balancing for the same horizons. Structural balancing on these horizons however had reveal another view showing that these mounded features were actually not vertically stacked as what we observed from horizon flattening technique. The mounded features turn out to be like a draping gentle fold, explaining that fault was synchronizing to displace these horizons during the time of deposition, suggesting a syn-depositional faulting. Flattened Horizon 5 give a dragging effect on Horizon 6 as highlighted in the rectangular of Figure 3D. However, this is not seen on the balanced section of Horizon 6 (Figure 3D').

4.0 Conclusion

It is good to practice either horizon flattening or structural balancing when performing seismic interpretation on a cross-section, because both techniques provide a mean to reveal the unseen features in the seismic data at the particular time of deposition. Horizon flattening is a simple, easy, user friendly and fast technique to help the interpreters predict what the situation during the time of deposition was. It can be very useful especially for a project that requires a quick-look interpretation result. Structural balancing or cross-section balancing/restoration, on the other hand, is more difficult and requires other information such as the rocks' densities, Poisson's Ratio and porosity values to get to a correct balancing result. However, with the availability of computer software such as MOVE software, these values are made available for testing purposes, and are good enough for work that involve less than 15km length of cross-section, like in our case. Instead of its difficulty compared to horizon flattening technique, structural balancing provide a better correct view for interpreting the environment of deposition and to understand what had happened to the sedimentation and faulting during the time of deposition. Despite all, both horizon flattening and structural balancing techniques are used throughout this work as procedures to enhance and validate our seismic interpretation. Generally, both techniques have their own benefits and drawbacks and what's important is, their geologically logic explanations.

References

[1] H. Fossen, *Structural geology*: Cambridge University Press, 2010.

[2] S. Bland, P. Griffiths, and D. Hodge. (2004) Restoring the seismic image with a geological rule base. *First Break*. 51-55.

[3] R. H. Groshong, "Cross Section Restoration and balancing as aid to seismic interpretation in extensional terranes," *3D Structural Geology,* vol. 1, pp. 299-318, 2002.

[4] M. G. Rowan and R. Kligfield, "Cross section restoration and balancing as aid to seismic interpretation in extensional terranes," *AAPG bulletin,* vol. 73, pp. 955-966, 1989.

Microdimensional pollution of the atmosphere of small settlements of the Far East of the Russian Federation

K.S. Golokhvast, V.V. Chaika, P.A. Nikiforov, Yu.S. Doroshev, N.V. Zemlyanaya, A.A. Fatkulin, V.P. Lushpey, Yu.A. Vasyanovich, A.M. Vasyanovich, A.I. Agoshkov

Far Eastern Federal University, 8 Sukhanova street, Vladivostok, 690950, Russian Federation

corresponding author: droopy@mail.ru

Abstract. Paper is devoted to research of atmospheric suspensions of the small cities and settlements of the Far East with the population to 100 thousand persons: Magadan, Belogorsk, Partizansk, Solovyevsk, Orotukan by means of methods of a laser granulometry and scanning electronic microscopy. Atmospheric suspensions were studied in the dropped-out snow which gathered in the different cities at the time of snowfalls from March, 2010 to January, 2013. It is shown that the studied settlements, considering dimension of fractions of atmospheric suspensions and their morphometric characteristics, it is possible to place in the following order on extent of increase of potential health hazard: Orotukan, Solovyevsk, Partizansk, Belogorsk, Magadan.

1. Introduction

The Far East is among the most interesting regions of Russia from the point of view of studying weighed in the atmosphere nano - and microparticles which, undoubtedly, have essential impact on quality of air, climate, people and animals. This interest is caused by a unique geographical arrangement of the cities, their remoteness from each other, low industrial activity and powerful seasonal transfers of air masses in ocean continent system.

Also long distances between the cities, a small amount of the large industrial enterprises and small number of the population are among interesting features of the Far East.

Earlier we investigated the large cities of the Far East: Vladivostok, Khabarovsk, Blagoveshchensk, Birobidzhan, Ussuriisk with the population from 100 to 700 thousand people [1].

This work is devoted to atmospheric suspensions of the small cities and settlements of the Far East with the population to 100 thousand persons: Magadan, Belogorsk, Partizansk, Solovyevsk, Orotukan.

2. Materials and methods

2.1. Samples area

Magadan - the city (an administrative center of the Magadan region) with the population of 95,9 thousand persons. It is located in the permafrost region, round hills on the bank of the Tauysky Bay

Sea of Okhotsk, on the isthmus connecting the peninsula to the continent between bays of Nagayev and Gertner. For Magadan the subarctic climate with lines of the sea is characteristic. Winter long, summer short and cool. The industry is founded on an energy drink and fishery. Selection was made in the downtown, near the highway.

Belogorsk - the city in the Amur region with the population of 75,16 thousand people. The city is located on the Tom River (inflow of Zeya). In Belogorsk sharp and continental climate with monsoonal lines and sharp prevalence of a summer precipitation. Winter cold, dry, with low-power snow cover. In the city of 7 boiler rooms (the downtown, to the area Amurselmash, to Gore's region) and Belogorsky shinoremontny plant. Selection at station was made around school No. 200 to Lenin St. address, 16. Nearby there is a highway and at distance about 1 km – the boiler room supplying with heat the downtown.

Partizansk - the city in Primorsky Krai with the population of 38,6 thousand people. It is located in a valley of the river Guerrilla. The city is surrounded by hills, their height generally 250 meters, some reach height of 500 meters. Climate of monsoonal type with warm damp summer and in the cold low-snow winter. Prior to the beginning of the end of the 1990th years Partizansk was the diversified industrial center, with prevalence of the coal industry. In 2008 coal mining in the village of Uglekamensk is stopped. In 10 km to the South from Partizansk, there is the Partizansky state district power station supplying with electricity all southeast of Primorye. The large railway station and locomotive depot Partizansk work. Tests of snow gathered in one point in the downtown.

Solovyevsk – the village (earlier the settlement) in the Amur region with the population of 3 thousand people. The large enterprises of the industry and others, except a boiler room, aren't present sources of dusting. Selection was made in the settlement center, near the highway.

Orotukan - the settlement of city type in the Magadan region with the population of 2,1 thousand people. Orotukan between medium-to-high hills is located on the right river bank. Orotukan's city-forming enterprise was the plant of the mountain equipment (OZGO) which existed from 1930th to 2005. Till 2002 in Orotukana there was an Orotukansky Mining and Processing Works (OMPW). For today large sources of dusting, except a boiler room, aren't available (www.orotukan.ru). Sampling was made in the settlement center, near the highway.

2.2. Methods

Atmospheric suspensions were studied in the dropped-out snow which gathered in the different cities at the time of snowfalls from March, 2010 to January, 2013. To exclude secondary pollution by anthropogenous aerosols, the top layer (5-10 cm) just dropped out snow was built. It placed in the 3 litre sterile containers. Through a couple of hours when snow in containers thawed, liquid shook up and from each sample gained 60 ml of liquid and analyzed on the laser analyzer of particles Fritsch Analysette 22 NanoTech (Germany). The analysis of samples carried out on a scanning microscope Hitachi S-3400N (Japan). Samples become covered by platinum.

3. Results and disscusion

Analyzing earlier received results [1], it is possible to divide aerosol particles by the sizes into seven classes: 1) from 0,1 to 1 micron (there corresponds PM1), 2) from 1 to 10 (there corresponds PM10), 3) from 10 to 50 microns, 4) from 50 to 100 microns, 5) from 100 to 400 microns, 6) from 400 to 700 microns and 7) more than 700 microns. The sizes of particles and percentage ratio of fractions in suspension tests in all settlements are specified in Table 1.

Table 1

Distribution of particles in snow on fractions in different settlements

<table>
<tr><th>∅, mkm</th><th>Magadan</th><th>Belogorsk</th><th>Partizansk</th><th>Solovyevsk</th><th>Orotukan</th></tr>
<tr><td>less then 1</td><td>-</td><td>-</td><td>-</td><td>-</td><td>-</td></tr>
<tr><td rowspan="4">1 - 10</td><td>2-3
13%</td><td>2-3
1%</td><td>2-3
2%</td><td>2-3
1%</td><td rowspan="2">7-9
3%</td></tr>
<tr><td>4-5
9%</td><td>4-6
7%</td><td rowspan="3">4-10
33%</td><td>4-7
7%</td></tr>
<tr><td rowspan="2">8-12
26%</td><td>5-7
6%</td><td rowspan="2">7-12
13%</td><td rowspan="2">9-12
9%</td></tr>
<tr><td>7-10
11%</td></tr>
<tr><td rowspan="3">10 - 50</td><td rowspan="3">12-15
52%</td><td>12-20
33%</td><td rowspan="3">20-40
30%</td><td>12-20
22%</td><td rowspan="3">15-18
13%</td></tr>
<tr><td rowspan="2">20-35
42%</td><td>25-30
8%</td></tr>
<tr><td>30-50
36%</td></tr>
<tr><td>50 - 100</td><td>-</td><td>-</td><td>50-100
35%</td><td>50-80
13%</td><td>70-100
3%</td></tr>
<tr><td>100 – 400</td><td>-</td><td>-</td><td>-</td><td>-</td><td>300-400
3%</td></tr>
<tr><td>400 - 700</td><td>-</td><td>-</td><td>-</td><td>-</td><td>-</td></tr>
<tr><td>more than 700</td><td>-</td><td>-</td><td>-</td><td>-</td><td>800-1000
69%</td></tr>
</table>

More detailed morphometric characteristics of particles of the suspension found in snow, are given below (table 2).

Table 2

Morphometric parameters of particles of the suspension containing in snow of different settlements

Parameters	Magadan	Belogorsk	Partizansk	Solovyevsk	Orotukan
Arithmetic Mean Diameter [μm]	9,68	18,84	34,12	28,98	607,93
Mode [μm]	12,56	26,39	66,76	39,71	1003,37
Median [μm]	11,01	17,50	27,45	27,88	920,10
Variance [μm^2]	13,61	67,26	729,08	289,96	195739,65
Mean Square Deviation [μm]	3,69	8,20	27,00	17,03	442,42
Coefficiant of Variation [%]	38,12	43,54	79,14	58,75	72,77
Spec.Surface Area [cm^2/cm^3]	8405,72	4326,28	4400,28	3462,35	1621,01

If to compare granulometric characteristics of suspensions in the studied settlements, it is possible to make some conclusions.

First, the qualitative structure of suspensions of all studied settlements is rather similar, and, judging by scanning electronic microscopy, for 80-90% consists of particles of natural minerals. It can do, is explained by that all these settlements are mainly on plains and in formation of suspensions in these territories the soil erosion is of great importance. Also It should be noted lack of large sources of dusting - combined heat and power plant and industrial facilities. The picture of qualitative structure of atmospheric suspensions typical for the small cities, on the example of Belogorsk, is given in figure 1.

Fig. 1. Typical suspension of small settlements (Belogorsk, winter 2011-2012). Scanning electronic microscopy in secondary electrons. Magnification x470.

Except natural components of suspensions, is, judging by morphology, and technogenic microparticles which condemn close attention of experts in the field of environmental protection. In tests practically all settlements in soot and a coal dust (according to the power dispersive analysis) microparticles of metals - Fe, by Zn, Cr, Cu, Ba are often found.

To be necessary to note rather enough high maintenance of suspensions belonging to the class PM10 in the studied cities and settlements (from 12% in Orotukana to 48% in Magadan). These particles from the point of view of environmental monitoring are considered dangerous, but considering their natural genesis (by results of scanning electronic microscopy), it should be taken into account them rather safe for health of the person.

It is important that despite the small sizes of particles (2 and the 3rd classes on our classification), they possess rather small area of a specific surface that besides testifies to rather safe ecological situation as it is considered that on a surface of particles various toxic substances are occluded. Therefore, the specific surface is higher, the danger of particles is higher.

4. Conclusions

As a result, the studied cities can be placed in the following order on extent of increase of microdimensional pollution of the atmosphere, according to morphometric and qualitative characteristics of particles of suspensions: Orotukan, Solovyevsk, Partizansk, Belogorsk, Magadan.

Work is supported by the Scientific Fund of FEFU and Grant President of Russian Federation for young scientists of MK-1547.2013.5.

References

[1] Golokhvast K.S. Atmospheric suspensions of the cities of the Far East. Vladivostok: FEFU Publishing house, 2013. 178 pages.

Structural Enhancement in Shallow Gas Cloud Region

A.H. Abdul Latiff[1], D.P. Ghosh, S.N.F. Jamaludin

Centre for Seismic Imaging, Universiti Teknologi PETRONAS, Malaysia
Email: abdulhalim.alatiff@petronas.com.my

Abstract: This research will address the issue of poor structural interpretation below complex overburden and shallow anomaly that cause wave field distortion. The main objective of this research paper is to present the result of seismic interpretation over a seismic line extracted from a shallow gas filed located in Malay Basin. The readers will be briefly explained on the seismic data properties of shallow gas cloud, followed by methodology and techniques used in interpreting the data. Interpretation techniques used in this research work include the seismic data analysis based on the seismic attributes. The outcome from this work aims at recovering the hidden structure that cannot be seen in the data after processing and imaging sequences.

1. Introduction

The shallow gas clouds is defined as an overburden region [1], and was formed due to the gas leak from deeper reservoir [2], migrating upward through the fault and fracture system. This thin layered gas-filled sediments that formed a shallow cloud was accumulated within porous silt and sand-rich sediments after a biological process which produced thermogenic gasses that not related to hydrocarbon such as carbon dioxide, hydrogen sulphide, ethane and methane [3]. Either way, the presence of shallow gas accumulation along the sediment layers in near surface, is always associated as the clearest indicator of the existence of hydrocarbon reservoir underneath the particular field, as proven in offshore Malaysia, North Sea and Gulf of Mexico. It was further elaborated that the shallow gas cloud can be recognised in seismic data based on several criteria [4]:

i. Acoustic turbidity – The smear reflector images caused by scattering energy. Adjacent reflector position will appear as a pull-down effect; indicated the delay in travel time throughout the slower zone.
ii. High amplitude reflection – High amplitude with negative phase reflection normally visible at the top of shallow gas zone. The greater amplitude reflected back from this area due to greater impedance contrast of gas-charged sediments compare to surrounding sediments.
iii. Acoustic blanking – This is the area where the reflection absent from the seismic image (**Figure 1**). The main caused of this 'wipe-out' was originated from wave scattering and amplitude attenuation.

---
[1] Corresponding author

In shallow gas cloud region, the acoustic blanking appears on the seismic data remains the main problem for seismic interpreters in order to locate the exact well location for drilling. This problem is attributed to non-uniform wave propagation from source to receiver location during the acquisition process. To enhance the structural image underneath the shallow gas cloud, kinematic (travel time delay) and dynamic (distortion of wave attributes) of the seismic data need to be corrected. The Diving Wave Tomography (DWT) technology [5] was used as one of the solution for reflector's travel time delay (kinematic) as well as the application of Pre-stack Depth Migration (PreSDM) technique that was extensively used in processing procedure. On the other hand, a dynamic approach is required to solve the frequency and amplitude losses as well as phase distortion. In standard data processing procedure, the amplitude loss and phase distortion were compensated by using a Q-compensation technique as well as wavelet sharpening and de-convolution processes. For example, a two steps approaches was introduced as kinematic aspect of seismic data was solved by using reflector's re-datuming concept while dynamic aspect of amplitude seismic data was solved through full waveform inversion (FWI) [6]. Each of the methods discussed will solve the travel time error and attenuation effect, although extensive computing power as well as sophisticated imaging algorithms will be needed.

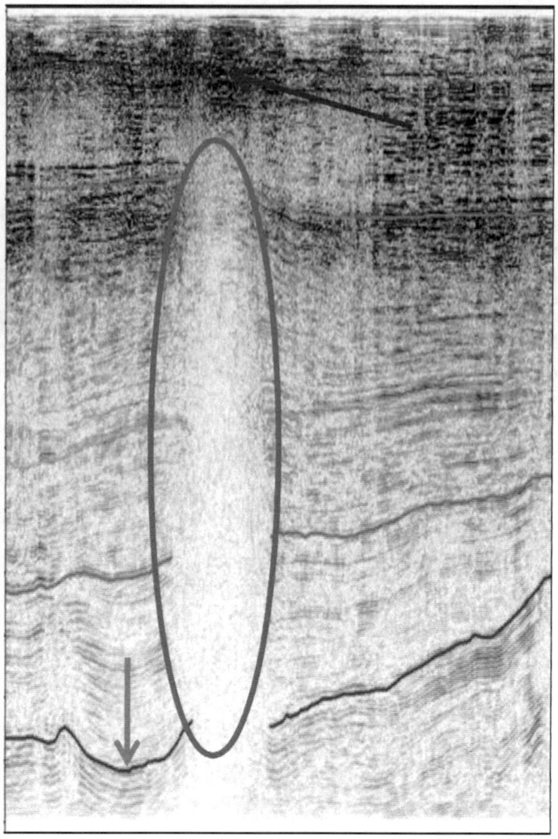

Figure 1: The shallow gas clouds zone was extracted from a 3D time migration seismic data, located in the Malay Basin. The red ellipse shows the vanished reflection events underneath the gas cloud, the green arrow shows the push-down effect appeared on the seismic data and the blue arrow indicated the strong amplitude reflection from the reflector located above shallow gas accumulation. (Data permission from PETRONAS).

2. Methodology

However, the resultant seismic section produced from these gas cloud region still requires the implementation of seismic attributes, for better structural enhancement, particularly within the 'wipe-out' zone. In definition, seismic attribute is a measure of seismic data that helps us to visually enhance interpretation interest which are sensitive to desired geological features that leads to identification of potential hydrocarbon reservoir by locating the true position of fault, fractures and well position through amplitude and geomorphological features characterization of the seismic section [7]. To date, many attributes have been extracted and developed for seismic interpretation purposes. To name some of them are volumetric curvature, instantaneous frequency, root mean square (RMS) amplitude, sweetness and many more. The most common attribute used for seismic interpretation is true amplitude attribute. The use of seismic attributes begins during 1980s and become more common in the 1990s.

Throughout this work, we incorporated several attributes to a seismic volume which suffered from poor illumination due to gas leakage. The main objective of attributes application is to enhance and improve our interpretation, especially when marking the faults, which is highly ambiguous using the present dataset. Three main seismic attributes (true amplitude, relative acoustic impedance and coherence-variance) were applied to the seismic volumes. These attributes are independently analysed in the next section along with the resultant seismic section and time slice produced after attribute implementation.

3. Results and Analysis

3.1 True Amplitude Attribute

The data true amplitude attribute was selected to display the original seismic trace. Seismic reflectivity can be displayed in variable density colour using seismic colour schemes available within the PETREL interpretation software (black-white or blue-red-white). Each of these colours represent positive (hard kicks) and negative (soft kicks). Horizons and structural interpretation were done based on true amplitude seismic at the beginning of the research, following the standard interpretation techniques by marking the horizons and faults visible on the seismic data.

From **Figure 2a**, interpretations of the horizons become more effortful at the deeper sections as amplitude of the reflectors dimmed and they experienced poor horizontal continuity. With the aim to improve the interpretation of the deeper reflectors beneath the shallow gas cloud area, relative acoustic impedance attribute has been applied to the seismic volume.

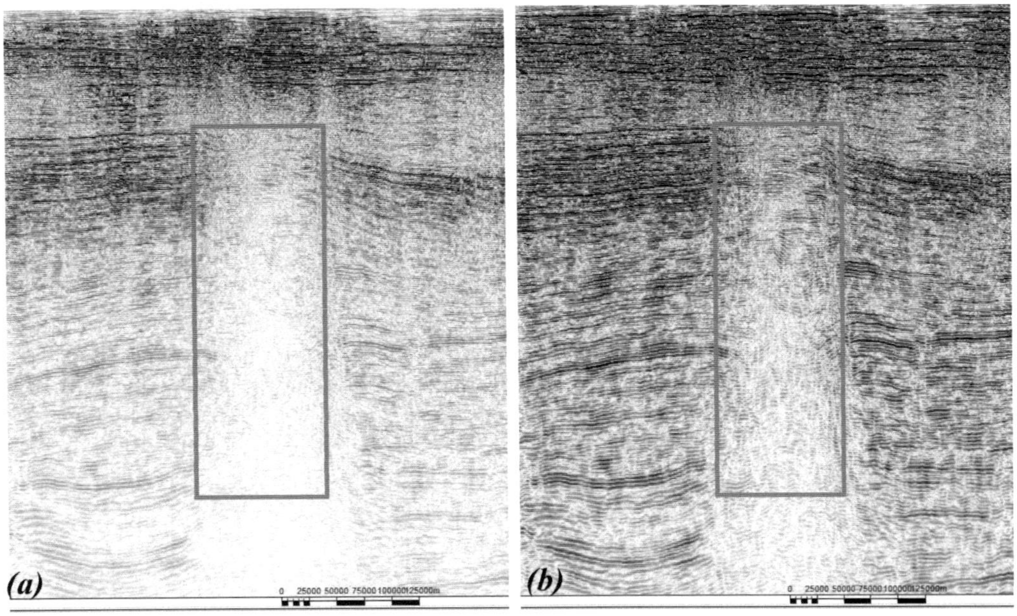

Figure 2: Comparison on a shallow gas cloud's cross-section in ***(a)*** true amplitude image and ***(b)*** relative acoustic impedance image. The true amplitude data is the original data obtained from seismic data acquisition without any changes to its amplitude spectrum. The reflectors within the wipe-out zone (highlighted in orange rectangle) are more visible on relative acoustic impedance attribute. (Data permission from PETRONAS).

IOP Conf. Series: Earth and Environmental Science **23** (2015) 012005 doi:10.1088/1755-1315/23/1/012005

3.2 Relative Acoustic Impedance

Relative acoustic impedance is a running sum of regularly sampled amplitude values. It is calculated by integrating the seismic trace, passing the result through a high-pass Butterworth filter, with a hard-coded cut-off at (10*sample rate) Hz. In simpler word, acoustic impedance is a physical rock property given as the product of density and interval velocity [7]. Application of acoustic impedance in the industry is usually by doing inversion of the seismic to get the correct values of the acoustic impedance. As that is not the aim of this research and due to time constraint, seismic inversion of acoustic impedance is not applied on the data. Result from application of relative acoustic impedance on the seismic cube is already sufficient to achieve the target of correlating the reflectors located just beneath the gas cloud zone, as shown in **Figure 2b.**

3.3 Coherence-Variance Attribute

Coherence is a measurement of similarity between waveforms and traces. It operates on a spatial window based on the neighbouring traces. Highly coherent seismic waveforms indicate homogeneity in the lithology, while abrupt changes in the waveform can indicate displacement of faults or fractures in the sediments. Coherence attribute provides assistance to seismic interpreters by portraying subtle faults or improve interpretation below the limits of seismic resolution. The main purposes of applying coherence attribute on the seismic cube are to recognize and improve the interpretation particularly for detecting faults underneath the shallow gas zone. The application of coherence attribute is best viewed in time slice domain and horizon surface.

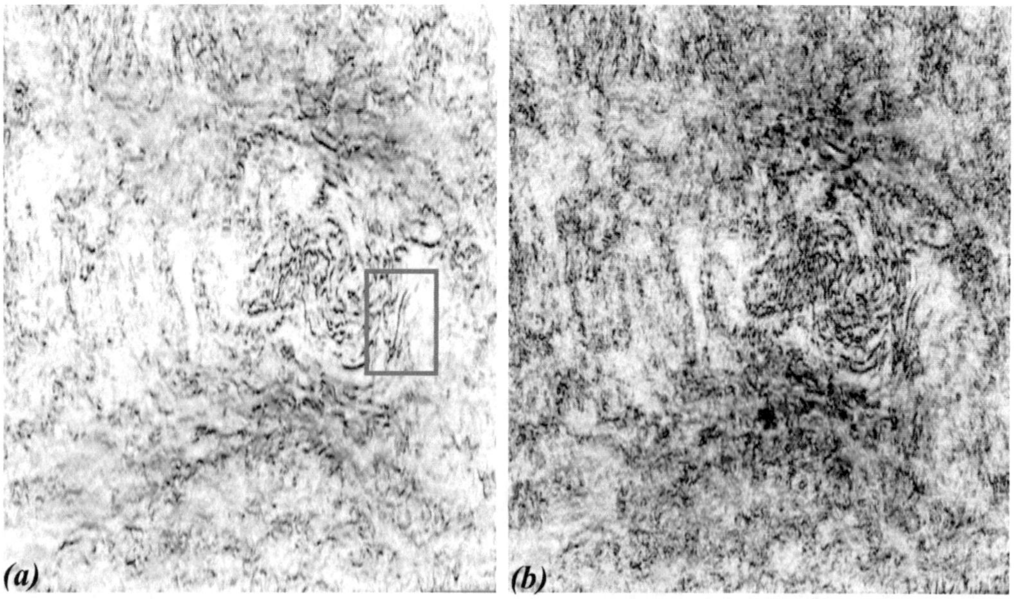

Figure 3: Comparison of time slices after application of coherence-variance attributes with: ***(a)*** application after structural smoothing and ***(b)*** application without structural smoothing (only dip and azimuth calculation). The implementation of structural smoothing to the seismic data before coherence attribute application increase the reflector visibility and continuity as seen in orange rectangle in ***(a)***. (Data permission from PETRONAS).

From several coherence calculations methods available, variance-based technique was used in this research. Variance method is the sum of the variance over a vertical analysis window of certain amount of samples (waveforms) and then normalize by the energy of all traces obtain from the seismic (Chopra and Marfurt [7]). A structural smoothing technique was applied prior to application of variance method. The purpose of applying structural smoothing on the input is to increase the continuity of the seismic reflectors using principal component of dip and azimuth to determine the local structure followed by structural smoothing, as demonstrated in **Figure 3**. Ignoring these procedures might introduce artefacts, hence misleading our interpretation.

Unlike in the area of higher resolution with better horizon continuity, the low amplitude distribution underneath the shallow gas cloud make it difficult to delineate fault system in this area. However, the implementation of coherence attributes on seismic section able to overcome this issue as demonstrated in **Figure 4a.** From the result, it clearly can be seen that a fault structure (indicate in red arrows) actually pass through the poor illumination zone (which is not visible in **Figure 1** and **Figure 2**). This aligns with the theory that the gas accumulation was probably formed due to gas leakage from deeper reservoirs through the fault system. Therefore the fault line interpreted in **Figure 4b** can only be verified through the application of coherence-variance attribute method.

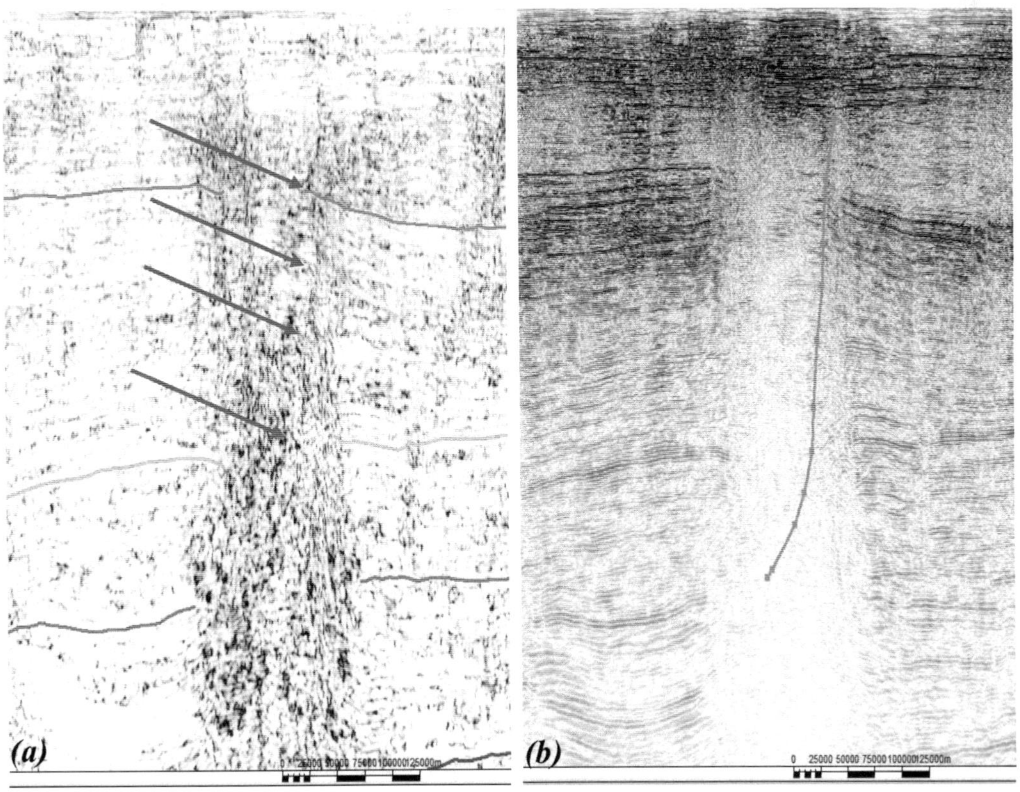

Figure 4: ***(a)*** Coherence attributes application after fault interpretation which the fault line clearly seen along the red arrows. ***(b)*** The same fault line then overlay into original (true amplitude) seismic section, give better insight of the fault within the wipe-out section. (Data permission from PETRONAS).

4. Conclusion

The application of seismic attributes in characterizing structure in poor illumination zone was shown throughout this work and it is proven as an effective tool for better fault and horizon detection for shallow gas cloud area. Without the implementation of seismic attributes to the seismic data, the fault detection process is extremely hard with no obvious fault line can be detected due to data 'wipe-out'. In this work, seismic interpretation becomes more effective and meaningful when more than one attributes were used and integrated to each other. Therefore the quality of the seismic data is paramount for delineating and characterizing reservoir beneath the gas cloud. However, we have to emphasize that in the shallow anomaly exploration, the input seismic data should be acquired and processed optimally, for example by employing advanced solutions in seismic acquisition and seismic data processing and imaging, as this is more critical than post-migration processing such as structural interpretation through seismic attributes method.

Acknowledgement

We would like to thank PETRONAS for financial support and permission to publish the data.

References

[1] R. E. Sheriff, *Encyclopedic dictionary of exploration geophysics*: Society of Exploration Geophysicists, 2002.
[2] B. Arnsten and M. Thompson, "The improtance of wide azimuth in imaging," in *EAGE 65th Conference & Exhibition*, Stavanger, Norway, 2003.
[3] G. D. Floodgate and A. G. Judd, "The origins of shallow gas," *Continental Shelf Research,* vol. 12, pp. 1145-1156, 1992.
[4] A. G. Judd and M. Hovland, "The evidence of shallow gas in marine sediments," *Continental Shelf Research,* vol. 12, pp. 1081-1095, 1992.
[5] M. C. Tanis, H. Shah, P. A. Watson, M. Harrison, S. Yang, L. Lu, *et al.*, "Diving wave refraction tomography and reflection tomography for velocity model building," presented at the SEG Annual Meeting, New Orleans, 2006.
[6] A. R. Ghazali, "True-amplitude seismic imaging beneath gas clouds," PhD, Applied Physics, Delft University of Technology, 2011.
[7] S. Chopra and K. J. Marfurt, *Seismic Attributes for Prospect Identification and Reservoir Characterization*: Society of Exploration Geophysicists, 2007.

Buckling Analysis of Rectangular Plates with Variable Thickness Resting on Elastic Foundation

K.K. Viswanathan[a,1], P.V. Navaneethakrishnan[b] and Z. A. Aziz[a]

[a]UTM Centre for Industrial and Applied Mathematics,
Department of Mathematical Sciences, Faculty of Science,
Universiti Teknologi Malaysia, 81310 Johor Bahru, Johor, Malaysia
[b]Vinayaka Missions University (Chennai office), Poonamalle High Road,
Chennai 600 010, India.

Abstract. Buckling of rectangular plates of variable thickness resting in elastic foundation is analysed using a quintic spline approximation technique. The thickness of the plate varies in the direction of one edge and the variations are assumed to be linear, exponential and sinusoidal. The plate is subjected to in plane load of two opposite edges. The buckling load and the mode shapes of buckling are computed from the eigenvalue problem that arises. Detailed parametric studies are made with different boundary conditions and the results are presented through the diagram and discussed.

1. Introduction

The buckling characteristics of non-uniform plates are utilized in many areas including aerospace engineering, engineering design and earthquake resistant structures. Significant studies have been made by '[1-3]' on buckling of plates of variable thickness. A spline function approximation was also used for such a study by '[4-7]' .The plate was homogenous with sinusoidal variation in thickness in the direction parallel to one edge. In the study of '[8]' a circular plate was considered with the plate thickness varying along the radius and the method of solution used a Frobenius series approximation. '[9]' studied on buckling loads of braced beam resting on elastic foundation using analytical method. However, not much information is available on buckling of plates of variable thickness resting on elastic foundation.

In the present work the buckling of thin rectangular plate of varying thickness, resting on elastic foundation is studied using spline function approximation technique. The plate is fully attached to the foundation. The plate thickness varies in the direction of an edge. Linear, exponential and sinusoidal variations in thickness are all considered in one formulation.

The solution of the differential equation characterising the deflection is assumed in a separable form. A closed form solution is generally not possible. A spline function numerical technique is used, exploiting the attractive features as excellent interpolating and approximation functions. '[10]' first presented it as an efficient tool to solve two-point boundary value problems, bringing out its computational superiority to other schemes like Hermite interpolation technique.

A pair of opposite edges of the rectangular plate are subjected to compressive uniform load. Two cases of boundary conditions are considered for these edges: Clamped-Clamped and Clamped- Simply supported. The other edges are simply supported. The deflection equation yields an eigenvalue problem solving which the critical loads and the modes shapes of buckling are obtained.

Parametric studies of the variation of the critical load with respect to aspect ratio, foundation coefficient and variation of thickness of the plate are made. Selected mode shapes of buckling are also presented.

[1] Corresponding author: K.K. Viswanathan, email: visu20@yahoo.com, viswanathan@utm.my

2. Formulation of the Problem

Consider a thin rectangular plate bounded by $x=0$, $x=a$, $y=0$ and $y=b$ as shown in 'figure 1'. The plate is subjected to an in-plane compressive force N_x acting on and normal to the edge $x=0$ and $x=a$. Its transverse deflection $w(x, y)$ is governed by the differential equation

$$\nabla^2(D\nabla^2 w)+(\upsilon-1)\lozenge^4(w,D)=q \tag{1}$$

where ∇^2 is Laplace's operator, υ is poisson ratio, $D=D(x,y)=Eh^3(x,y)/12(1-\upsilon)^2$ is the variable flexural rigidity, $h=h(x,y)$ is the variable thickness, E is Young's modulus, q is the distributed normal load and

$$\lozenge^4(D,w)=\frac{\partial^2 D}{\partial x^2}\frac{\partial^2 w}{\partial y^2}-2\frac{\partial^2 D}{\partial x\partial y}\frac{\partial^2 w}{\partial x\partial y}+\frac{\partial^2 D}{\partial y^2}\frac{\partial^2 w}{\partial x^2} \tag{2}$$

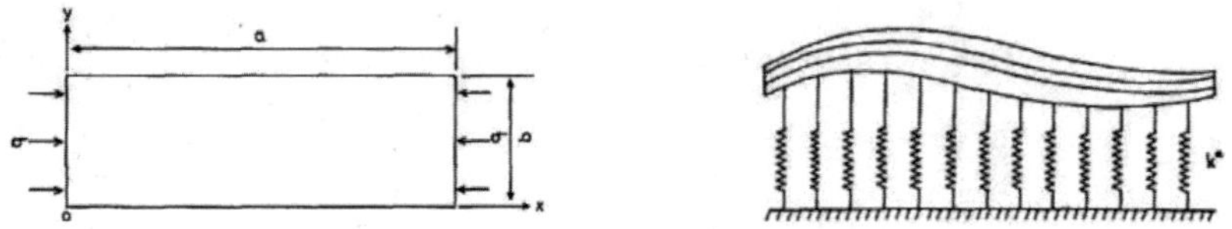

a) Distributed Normal Load $=q_x$ b) Foundation Modulus $=k^*$

Figure 1 Rectangular plate of variable thickness on elastic foundation a) Geometry and load b) Description of elasticfoundation

The plate is fully attached to the elastic foundation of elastic coefficient k^*. The intensity of reaction is proportional to k^*w. Then

$$q_x=-N_x\frac{\partial^2 w}{\partial x^2}-K^*w-\rho h\frac{\partial^2 w}{\partial t^2} \tag{3}$$

The vibration of the plate is not considered and hence the last term of 'equation (3)' will be absent. Let the edge $y=0$ and $y=b$ of the plate be simply supported, then the transverse deflection can be assumed in the separable form

$$w(x,y)=\overline{W}(x)\sin\frac{n\pi y}{b} \tag{4}$$

'Equation (2-4)' are used in 'equation (1)'. The thickness of the plate is assumed as a function of x only, given by

$$h(x)=h_o\left[1+C_\ell\frac{x}{a}+C_e e^{x/a}+C_s\sin\frac{n\pi}{a}\right] \tag{5}$$

If C_ℓ, C_e and C_s are all zero, then the plate thickness becomes constant. If C_e and C_s are zero, then the thickness varies linearly. If C_ℓ and C_s vanish then the thickness variation is exponential. If C_ℓ and C_e are zero then the thickness variation is sinusoidal. We impose the conditions $C_\ell>-1$, $C_e<-1/e$ and $C_s>-1$, which ensure that the thickness is always positive.

Assume also

$$N_x=\sigma h_o,\quad X=\frac{x}{a},\quad Y=\frac{y}{b},\quad W=W(x)=\frac{\overline{W}}{a},\quad H=\frac{h_o}{a},\quad \phi=\frac{q}{b},\quad f=n\phi\pi \tag{6}$$

A buckling coefficient K and a foundation coefficient L are defined by

$$K = \frac{12(1-\upsilon^2)\sigma}{\pi^2 EH^2\phi^2}, \qquad L = \frac{k^* a}{EH^3} \tag{7}$$

For the least critical value of K, n is least '[11]' and hence we set $n = 1$. The differential 'equation (1)' becomes

$$W^{iv}(x) + A_0 W''' + A_1 W'' + A_2 W' + A_3 W = 0 \tag{8}$$

where

$$A_0 = \frac{6(C_\ell + C_e e^x + C_s \pi q)}{(1 + C_\ell X + C_e e^x + C_s p)}$$

$$A_1 = \frac{6(C_\ell + C_e e^x + C_s \pi q)^2}{(1 + C_\ell X + C_e e^x + C_s p)^2} + \frac{3(C_e e^x - C_s \pi^2 p)}{(1 + C_\ell X + C_e e^x + C_s p)} - 2f^2 + \frac{Kf^2}{(1 + C_\ell X + C_e e^x + C_s p)^3}$$

$$A_2 = \frac{6f^2(C_\ell + C_e e^x + C_s \pi q)}{(1 + C_\ell X + C_e e^x + C_s p)}$$

$$A_3 = f^4 - 3\upsilon f^2 \left\{ \frac{2(C_\ell + C_e e^x + C_s \pi q)^2}{(1 + C_\ell X + C_e e^x + C_s p)} + \frac{3(C_e e^x - C_s \pi^2 p)}{(1 + C_\ell X + C_e e^x + C_s p)} \right\} + \frac{12(1-\upsilon^2)L}{(1 + C_\ell X + C_e e^x + C_s p)}$$

(9)

$$p = \sin \pi X \quad \text{and} \quad q = \cos \pi X \tag{10}$$

The 'equation (8)' is a linear differential equation with variable coefficients. Boundary conditions are imposed on the edges, $X = 0$ and $X = 1$. Thus the 'equation (8)' constitutes as well-defined boundary value problem for $0 \le X \le 1$.

3 Method of Solution

Since a closed form solution for the differential 'equation (8)' does not exist, we hence to resort to numerical methods for solving it. The collocation procedure, using a quintic spline approximation for $W(X)$ is used because the 'equation (8)' is of order four.

Therefore, the approximating spline function is assume to be

$$W^*(X) = \sum_{i=0}^{4} a_i X^i + \sum_{j=0}^{N-1} b_j (X - X_j)^5 H(X - X_j) \tag{11}$$

In which $H(X - X_j)$ is the Heaviside function and a_i, b_j are the unknown coefficients.

This function has very attractive characteristics compared to many others from the point of view of convergence, accuracy and elegance of usage on computer and efficiency for computational work.

Let us assume that the subintervals are all equal in length. The knots are at

$$X = X_s = \frac{S}{N}, \qquad s = 0, 1, 2, \ldots, N.$$

Imposing the condition that the 'equation (11)' satisfies 'equation (8)' at each $X = X_s$, we have the following system of $N+1$ equations in the $N+5$ spline coefficients : a_0, a_1, a_2, a_3, a_4, b_0, b_1, …, b_{N-1}.

4 Boundary conditions

The boundary conditions of the plate considered are as follows:

i. (C – C) : Clamped at both the edges $X = 0$ and $X = 1$
ii. (C – S) : Clamped at $X = 0$ and Simply supported at $X = 1$

The pairs of boundary conditions that should be satisfied at a clamped edge, a simply supported edge and a free edge are respectively,

$$w = \frac{\partial w}{\partial x} = 0 \,(\text{Clamped}), \quad w = \frac{\partial^2 w}{\partial x^2} = 0 \,(\text{simply supported}) \tag{12}$$

Each of these cases gives four more equations on the spline coefficients. Combining these equations with 'equations (8)' we have a generalized eigenvalue problem in the form of

$$[M]\{c\} = K[P]\{c\} \tag{13}$$

Where $[M]$ and $[P]$ are the square matrices of order $(N+5)$ and $\{c\}$ is a column vector whose elements are the spline coefficients. The eigenparameter is the buckling coefficient K, whose least value only is of interest since buckling occurs for the corresponding minimum value of σ, the critical load. After obtaining an initial approximation to the least value of K, its finer approximation is obtained using Regular-Falsi method. The eigenvector $\{c\}$ corresponding to this least K determines the spline function $W^*(X)$ from which the mode shapes of buckling of the plate can be constructed.

5 Results and Discussion

'Table 1' shows the convergence study on K with respect to the number of knot $N = 6(2)18$. Consistent improvement was observed in the value of K with the increase in the number of knots. It has been found that it sufficient to keep $N = 12$ since further increase in N improves K by less than 0.16%.

Table 1 Convergence study

N	$C_e = -0.2, L = 1000, \phi = 0.5$		$C_\ell = -0.5, L = 200, \phi = 0.5$	
	C-S		C-C	
	K	% change	K	% change
6	37.70713	-	29.26319	-
8	37.83680	0.3439	29.33973	0.2410
10	37.90233	0.1732	29.38955	0.1698
12	37.96188	0.1571	29.43378	0.1505
14	38.01875	0.1498	29.47737	0.1481
16	38.06821	0.1301	29.51858	0.1398
18	38.12082	0.1382	29.55642	0.1282

The $K-\phi$ relations obtained for $L=0$ (not attached foundation) and $C_\ell = C_e = 0$ (sinusoidal thickness variation) are seen to completely agree with the results of '[12]'. The $K-\phi$ relations obtained for $L=0$ and $C_\ell = C_e = C_s = 0$ (constant thickness) are matching perfectly with those of '[13]'. The other results are new.

The following parametric values are considered to analyse the problem.

$$\begin{aligned} &\upsilon = 0.3,\ \phi = 0.5(0.25)2.5, C_\ell = -0.5(0.25)0.5, C_e = -0.2(0.1)0.2, \\ &C_s = -0.5(0.25)0.5, L = 0(200)1000 \end{aligned} \tag{16}$$

The nature of variation of the least value of the buckling coefficient K with respect to each of the five variable parametric values $(\phi, C_\ell, C_e, C_s$ and $L)$, with the other four parametric values being fixed, is studied and depicted in the 'figures (2-4)'. Some typical mode shapes of buckling of the plate are presented in 'figure 5'.

The rate of change of K with ϕ may be discontinuous in some cases when $L=0$ ('[2]'and '[12]'). It is now found to be so in some cases even when $L \neq 0$.

'Figures (2-4)' depict the nature of variations of the value of the buckling coefficient K with respect to the increase of the aspect ratio ϕ for plates whose thickness varies linearly $(C_e = C_s = 0)$, exponentially $(C_s = C_\ell = 0)$ and sinusoidally $(C_\ell = C_e = 0)$ for different values of the thickness variation parameter C_ℓ, C_e and C_s respectively, for two fixed values of the foundation coefficient, $L = 200$ and $L = 1000$. They show clearly that the value of K in general reduces with the increasing ϕ for fixed values of the other parameters for all the type of boundary conditions considered. Again the K - values are found to be higher for higher values of the thickness variation parameters C_ℓ, C_e and C_s. Further K is found to be considerably higher for the higher value of the foundation coefficient $(L = 1000)$ than for its lower value $(L = 200)$. These can be clearly attributed to the facts that the stiffness of the plate decreases with increasing ϕ and increases with C_ℓ, C_e, C_s and L. It is also seen from the same figures that the K values are the least, in general, for the C-S boundary conditions and a higher for the C-C boundary conditions.

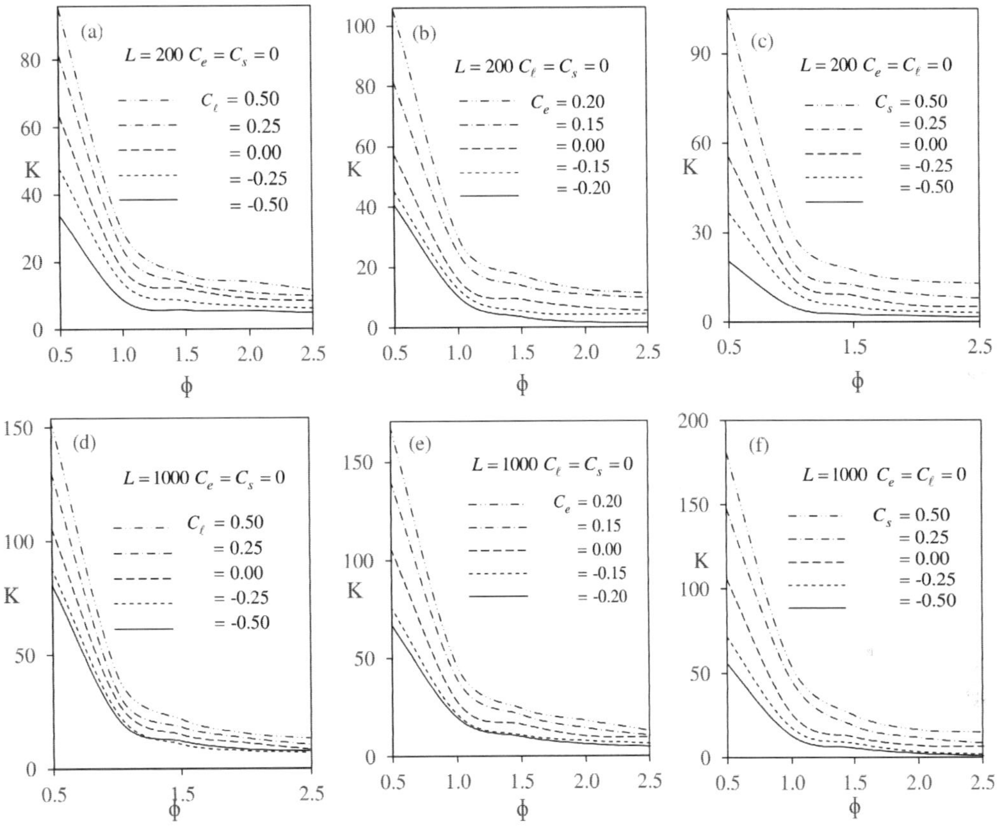

Figure 2 $K - \phi$ Diagrams for C-C plates

'Figure 4' presents the nature of change of K with respect to L for two cases of aspect ratio, $\phi = 1.5$ and $\phi = 2$ and the C-S boundary condition. As before, the linear $(C_e = C_s = 0)$, exponential $(C_s = C_\ell = 0)$ and sinusoidal $(C_\ell = C_e = 0)$ variations in thickness are considered. With increasing L, the values of K are seen, in all cases, to increase, almost steadily for long ranges of values of L. The other characteristics of variation of K agree with those studied from 'figures (2-4)'.

Qualitatively similar characteristics are observed in the variation of K with L for the considered other boundary conditions also. Some mode shapes of buckling are presented in 'figure 5'. They can be viewed as the buckled shapes of the central line $y = b/2$ of the rectangular plate considered. The two rows of diagrams correspond to the two types of boundary conditions: C-C and C-S. The two columns corresponds to the three types of thickness variation: linear (three cases), exponential (two cases) and sinusoidal (two cases). The case 3 of diagrams in (a), (d) and (g) corresponds to constant thickness $(C_\ell = C_e = C_s = 0)$ and therefore is not shown in the other diagram to avoid repetition.

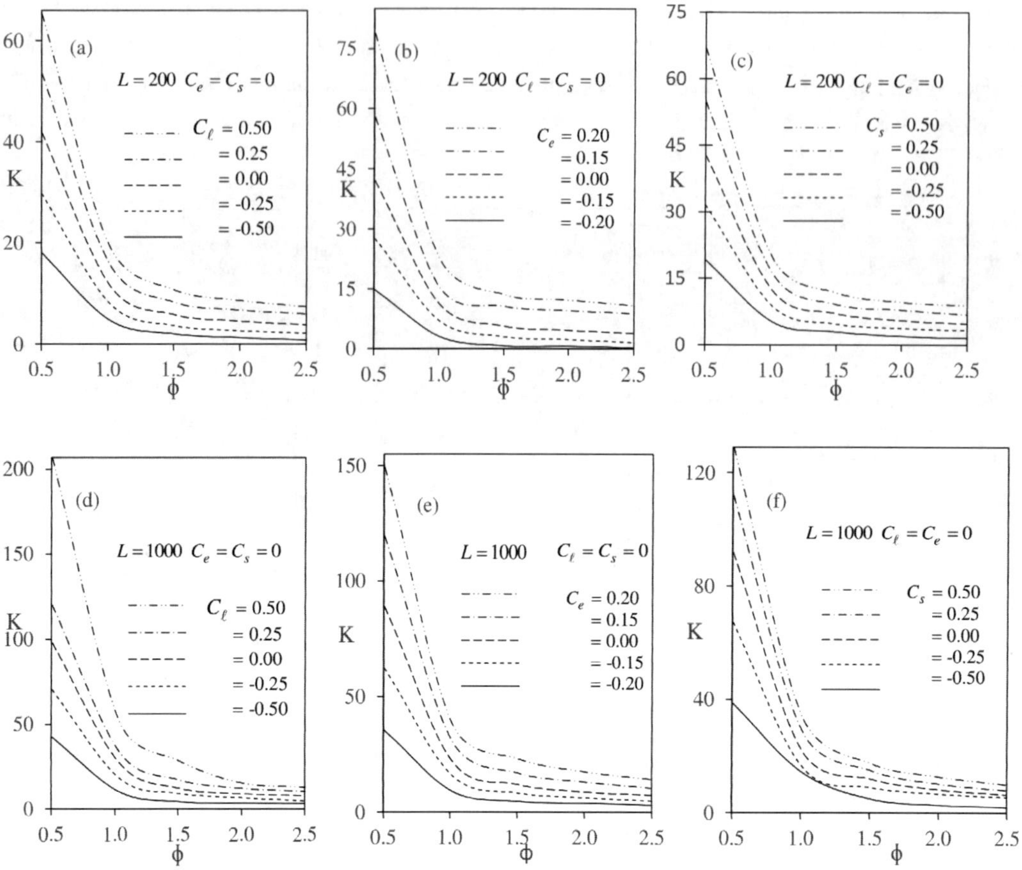

Figure 3 $K - \phi$ Diagrams for C-S plates

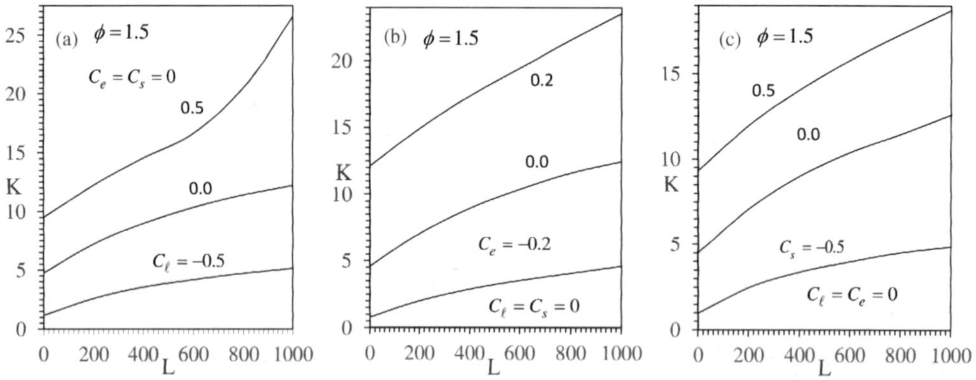

Figure 4 $K - L$ Diagrams for C-S plates

Two types of foundations are considered. The plate is not a square with $\phi = 0.5$. The deflections have been normalized with respect to the numerically largest deflections. It can be observed that the two mode shapes marked 3 in 'figure 5 (a)' and all mode shapes in 'figure 5 (c)' are either symmetric or antisymmetric since they corresponds to symmetric geometric and boundary conditions (plates of constant thickness and sinusoidally varying thickness, under C-C conditions). The mode shapes in general, are interesting and informative. The buckling mode numbers for corresponding buckling loads, for example, can be easily inferred from the mode shapes.

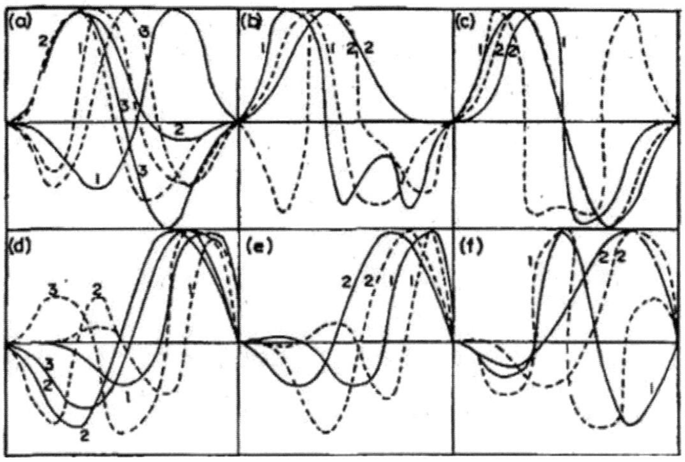

Figure 5 Mode shapes of Buckling.

Boundary conditions: (a), (b), (c): C-C, (d), (e), (f): C-S.

Thickness Variation : (a), (d): $C_e = C_s = 0$; $1: C_\ell = 0.5$; $2: C_\ell = -0.5$; $3: C_\ell = 0.0$

(b), (e): $C_s = C_\ell = 0$; $1: C_e = -0.2$; $2: C_e = 0.2$.

(c), (f): $C_\ell = C_e = 0$; $1: C_s = 0.5$; 2: $C_s = 0.5$.

Foundation Modulus: ——: $L = 200$; - - - - : $L = 1000$.

6 Conclusion

Buckling of thin plates of linearly, exponentially and sinusoidally varying thickness and subject to a constant in plane load has been studied. The plate is resting on an elastic foundation and is subject to two types of boundary conditions on a pair of opposite edges with the other edges simply are supported. An elegant spline function approximation for the unknown transverse deflection is used and the buckling loads and mode shapes of buckling are computed. Extensive parametric studies are made.

7 Acknowledgment

The authors thankfully acknowledge the financial support from the UTM-Flagship Project with the Vote. No. 01G40 under Research Management Centre (RMC), Universiti Teknologi Malaysia for completion of this research work and to present in Aeroearth 2014 Conference.

REFERENCES

[1] Olson R G 1939 Beitrag zur knicking der rechteckplate von quadratisch veranderlicher Veranderlicher steifigkeit. *Ing. Archiv.* **10** 175-181

[2] Wittrick W H and Ellen C H 1962 Buckling of Tapered Rectangular Plates in Compression. *Aeronautical Quarterly*. **3(4)** 302-326

[3] Hwang S S 1973 Stability of Plates with Piecewise Varying Thickness. *Journal of Applied Mechanics*. **40(4)** 1127-1129

[4] Viswanathan K K, Zainal A A, Amirah H Z and Saira Javed 2014 Free Vibration of Symmetric Angle-Ply Circular Cylindrical Shells. *IOP Conf. Series: Earth and Environmental Science*. **19** 1-10

[5] Viswanathan K K, Farah Syazwan M S, Mohamad M N, Saira Javed and Lee J H 2013 Free Vibration of Symmetric Angle-Ply Laminated Annular Circular Plates. *Int. J. Engg. & Tech.* **5(4)** 3554-3569

[6] Viswanathan K K, Saira Javed and Zainal A A 2013 Free Vibration of Symmetric Angle-Ply Laminated Conical Shell Frusta of Variable Thickness including Shear Deformation Theory by Spline Method. *Structural Engineering & Mechanics, An International Journal*. **45(2)** 259-275

[7] Viswanathan K K, Lee J H, Zainal A A and Hossain I 2011 Free Vibration of Symmetric Angle-Ply Laminated Cylindrical Shells of Variable Thickness. *Acta Mechanica*. **221** 309-319

[8] Gupta U S and Lal R 1978 Axisymmetric Vibrations of Linearly Tapered Annular Plates Resting on Elastic Foundation by Spline Method of Solution. *Journal of the Aeronautical Society of India*. **31** 97-102

[9] Zhang Y, Liu Y, Chen P and Murphy K D 2011 Buckling Loads and Eigen Frequencies of a Braced Beam Resting on an Elastic Foundation. *Acta Mechanica Solida Silica*. **24** 510-518

[10] Bickley W G 1968 Piecewise Cubic Interpolation and Two-Point Boundary Problems. *Computer Journal*. **11** 206-208

[11] Leissa A W 1969 *Vibration of Plates NASA SP-160* (Washington DC)

[12] Navaneethakrishnan P V 1988 Buckling of Non-Uniform Plates: A Spline Method. *Journal of Engineering Mechanics Division ASCE*. 114(5) 97-102

[13] Bulson P S 1970 *The Stability of Flat Plates* (Chatto and Windus : London) 58

IOP Conf. Series: Earth and Environmental Science **23** (2015) 012007 doi:10.1088/1755-1315/23/1/012007

A shallow water model for the propagation of tsunami via Lattice Boltzmann method

Sara Zergani, Z. A. Aziz and K. K. Viswanathan[1]
UTM Centre for Industrial and Applied Mathematics,
Department of Mathematical Sciences, Faculty of Science,
Universiti Teknologi Malaysia, 81310 Johor Bahru, Johor, Malaysia.

Abstract. An efficient implementation of the lattice Boltzmann method (LBM) for the numerical simulation of the propagation of long ocean waves (e.g. tsunami), based on the nonlinear shallow water (NSW) wave equation is presented. The LBM is an alternative numerical procedure for the description of incompressible hydrodynamics and has the potential to serve as an efficient solver for incompressible flows in complex geometries. This work proposes the NSW equations for the irrotational surface waves in the case of complex bottom elevation. In recent time, equation involving shallow water is the current norm in modelling tsunami operations which include the propagation zone estimation. Several test-cases are presented to verify our model. Some implications to tsunami wave modelling are also discussed. Numerical results are found to be in excellent agreement with theory.

1. Introduction

General patterns and important characteristics of tsunami can be predicted by various sets of governing equations. Euler equations remain the foundation basis for tsunami propagation models, it describes the frictionless motion of water under the influence of gravity, and a number of equations that result from them in asymptotic limits '[1]', including the classical wave equation, the forced Korteweg-de Vries (fKdV) equation, the Bousinnesq equation, or the shallow-water equations. This theory deals with the gravity waves problem in a regular and arbitrary depth ocean, from which tsunami wave solution is derivable from the shallow water wave approximation to its full solution. Nonlinearity plays an important part in transforming tsunami waves in the region. These difficulties can be solved by nesting near-field models like as the full Boussinesq equation model or modelling of shallow water equation with a finer mesh system to the present transoceanic model as described in '[2]'. The NSW equations are accurate for long wave propagation and runup problems, in which the scale of the vertical length scale is smaller than that of the scale of the horizontal length, such as for most tsunamis. Besides tsunami, these equations are widely used in the field of ocean engineering (e.g., tide and ocean modelling) and atmospheric modelling, where one length scale is dominant. This work focuses on the modelling of tsunami using the lattice Boltzmann method or lattice Boltzmann model for shallow water equations (LABSWE). This technique has been selected due largely to its computationally efficient basic lattice Boltzmann algorithm, and its capability to handle complex geometries and topologies. Several works have already applied LBMs to standard shallow water benchmark problems and test cases. '[3]' presented the so-called D2Q9 LBM implementation for the simulation of wave runup on a sloping beach. '[4]' applied a similar LBM to test cases including bed slope and friction terms. The main focus of their work was to demonstrate the ability of the technique to cope with multifaceted geometries and random bathymetry. Although LBMs are accepted generally to deal with complex geometries and interfacial dynamics, certain difficulties emerge in the case of boundary conditions.

2 Formulation of the problem

2.1 Governing equations

Since the scale of the vertical length is lesser than the horizontal length (refer Figure1), in order to achieve shallow water equation, both the continuity and Navier-Stokes equations have to be integrated in-depth. Shallow

[1] Corresponding author : K.K. Viswanathan, email: visu20@yahoo.com, viswanathan@utm.my

water equation are obtained from the depth integral of mass transport equation. The Coriolis effect for shallow water equations having forced wind term, bed slope and bottom friction terms is expressed as in '[5]':

Continuity Equation: $\frac{dh}{dt}+\frac{d(hu_i)}{dx_i}=0$ (1)

Momentum Equations: $\frac{d(hu_i)}{dt}+\frac{d(hu_iu_j)}{dx_j}+\frac{d}{dx_i}\left(\frac{gh^2}{2}\right)=\gamma\left[\frac{\partial^2(hu_i)}{dx_ix_j}\right]+F_i$ (2)

where i and j denote Cartesian indices and Einstein summation respectively; h is depth of the water, u_i is the i th direction-average depth velocity component, t denotes time. The term relating to forces is expressed as:

$$F_i = F_{pi}+F_{bi}+F_{wi}+F_{ci}, \quad F_{pi}=-gh\frac{\partial z_b}{\partial x_i}, \quad F_{bi}=C_bu_i\sqrt{u_iu_i}, \quad F_{wi}=\frac{\rho_a}{\rho_w}C_wu_{wi}\sqrt{u_{wi}u_{wi}}, \quad F_{ci}=\begin{cases} f_chu_y i=x \\ -f_chu_x i=y \end{cases},$$

where z_b is bed elevation, $g=9.81m/s^2$ acceleration force due to gravity, γ viscosity of the kinematic energy and F_{pi}, is the force from outside exacted on the shallow water flow that consists of an appropriate hydrostatic pressure approximation, F_{wi} denotes wind shear stress, F_{bi} is the bed shear stress, and the Colioris effect forcing term, F_{ci}. $F_c=2\omega\sin\varphi$ is the Coriolis parameter and ω is rotation rate of the earth and φ is the latitude, z_b is the bed elevation, $C_b=g/c_z^2$ is the bed friction coefficient and $C_z=h^{\frac{1}{6}}/n_b$ is both the Chezy and Manning coefficients at the bed, n_b, $\left[L^{-\frac{1}{3}}T\right]$.$\rho_\omega$ is the water density, ρ_a the air density, $C_w=\left(0.63+0.66\sqrt{u_{wi}u_{wi}}\times10^{-3}\right)$ is the expression for the coefficient of the wind, and u_{wi} is the i th direction velocity of wind.

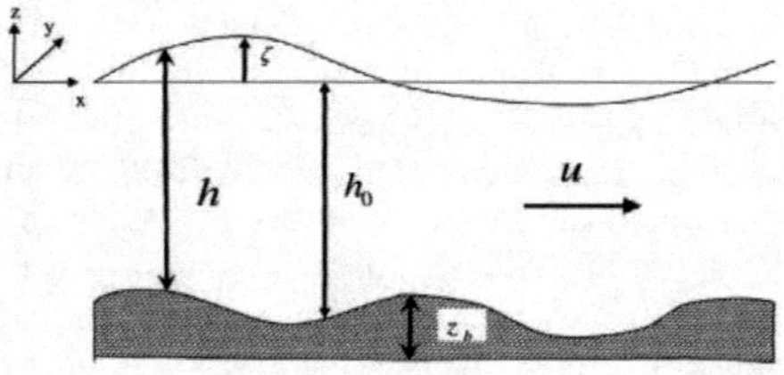

Figure 1 Shallow water flow regime

2.2 Lattice Boltzmann equation

The LBM is a numerical method for solution of flow equations without using the complicated shallow-water equations. It solves the lattice Boltzmann equation, and the depth and velocity can be calculated from macroscopic properties. Only simple arithmetic calculations are required to generate accurate solutions to flow problems with straightforward treatment of boundary conditions, and providing an easy and efficient way to simulate complicated flows.The lattice Boltzmann equation which includes a force term on a nine-velocity square lattice is given by

$f_\alpha(X+e_\alpha\Delta t,t+\Delta t)-f_\alpha(X,t)=-\frac{1}{\tau}\left(f_\alpha-f_\alpha^{eq}\right)+\frac{\Delta t}{6e^2}e_{\alpha i}F_i$ (3)

where $f_\alpha(x,t)$ is particle distribution function, x is space vector, t is time, e_α is particle velocity vector, where $\alpha=1,...,9$, $e=\Delta x/\Delta t$, Δx is lattice size, Δx is time step, τ is single relaxation time factor. The stability of the equation requires that $\tau>1/2$ and $f_\alpha^{(eq)}(x,t)$ is the equilibrium distribution function at time t. F_i is the i th direction force component.

2.3 Definition of macroscopic quantity

The water depth h is given as $h(X,t)=\Sigma_\alpha f_\alpha(X,t)=\Sigma_\alpha f_\alpha^{eq}(X,t)$ (4)

The macroscopic quantity velocity $u(X,t)$ is defined as

$$u_i(X,t)=\frac{1}{h(X,t)}\Sigma_\alpha e_{\alpha i} f_\alpha(X,t)=\frac{1}{h(X,t)}\Sigma_\alpha e_{\alpha i} f_\alpha^{eq}(X,t) \tag{5}$$

$$\Sigma_\alpha e_{\alpha i} e_{\alpha j} f_\alpha^{eq}(X,t)=\frac{1}{2}gh^2(X,t)\delta_{ij}+h(X,t)u_i(X,t)u_j(X,t) \tag{6}$$

2.4 Equilibrium Distribution Function For 2D Shallow Water Equations

Considering the theory of the lattice gas automata, the equilibrium function is the Maxwell-Boltzmann equilibrium distribution function. This distribution function is often expanded as a Taylor series in macroscopic velocity to its second order. It is assumed that an equilibrium function can be stated as a power series in macroscopic velocity

$$f_i^{eq}=A_\alpha+B_\alpha e_{\alpha i}u_i+C_\alpha e_{\alpha i}e_{\alpha j}u_iu_j+D_\alpha u_iu_j \quad , \tag{7}$$

It is convenient to write the equation above in the following form,

$$f_\alpha^{eq}=\begin{cases} A_0+D_0u_iu_i\alpha=0 \\ A+Be_{\alpha i}+Ce_{\alpha i}e_{\alpha j}u_iu_j+\bar{D}u_iu_i\alpha=1,3,5,7 \\ A+Be_{\alpha i}u_i+Ce_{\alpha i}e_{\alpha j}u_iu_j+Du_iu_j\alpha=2,4,6,8 \end{cases} \quad . \tag{8}$$

The coefficients can be determined based on the limitations of the equilibrium distribution function. The macroscopic quantity's three conditions must be satisfied by the local equilibrium distribution function in shallow water equation. The calculation of the Lattice Boltzmann equation leads to the resolution of the 2D equation for shallow water if the local equilibrium function could be determined under the above constraint. Substituting 'equation (8)' in 'equations (4-6)' and evaluating the terms with

$$e_\alpha=\begin{cases} (0,0)\alpha=0 \\ e\left[\cos\frac{(\alpha-1)}{4},\sin\frac{(\alpha-1)\pi}{4}\right], \ \alpha=1,3,5,7 \\ \sqrt{2}\left[\cos\frac{(\alpha-1)\pi}{4},\sin\frac{(\alpha-1)\pi}{4}\right], \ \alpha=2,4,6,8 \end{cases} \tag{9}$$

and after the coefficients are decided, this results in

$$f_\alpha^{eq}=\begin{cases} h-\frac{5gh^2}{6e^2}-\frac{2hu_iu_i}{3e^2} & \alpha=0 \\ \frac{gg^2}{6e^2}+\frac{he_{\alpha i}u_i}{3e^2}+\frac{he_{\alpha i}e_{\alpha j}u_iu_j}{2e^4}-\frac{hu_iu_i}{6e^2} & \alpha=1,3,5,7 \\ \frac{gh^2}{24e^2}+\frac{he_{\alpha i}u_i}{12e^2}+\frac{he_{\alpha i}e_{\alpha j}u_iu_j}{8e^4}-\frac{hu_iu_i}{24e^2} & \alpha=2,4,6,8 \end{cases} \tag{10}$$

2.5 Boundary and initial conditions

In order to solve shallow water flow problems by use of LABSWE, suitable boundary conditions must be provided. Generally speaking, in the application of boundary conditions in LBM, the temporal/spatial flexibility is allowed. This is briefly described as follows: solid boundary conditions; no-slip or slip boundary conditions may be used. For no-slip conditions, the normal bounce-back scheme can be applied. For slip conditions, a zero gradient of the distribution function perpendicular to the solid wall can be employed. Representation of boundary inflow and outflow and periodic boundary conditions are used in the verification of the models.

3 Results and Discussion

In the following, the NSW-LBM code is applied to three benchmark problems or test-cases widely used in the tsunami community: (i) 2D Tidal flow over a regular bed in 3D plot; and (ii) 2D Tidal flow over an irregular bed in 3D plot; and (iii) 2D steady flow over an irregular bed in 3D plot. In all cases, the LBM solution is compared to the available analytical/numerical or experimental reference solutions, and to other results from the literature.

3.1 2D Tidal flow over a regular bed in 3D plot

Tidal waves often occur in coastal engineering. '[6]' test problem is used in verifiying upwind discretisation of the source of bed slope terms. A two dimensional problem in which is defined the bed topography as $G(x)=50.5-40x/L-10\sin\left[\pi(4t/L-1/2)\right]$,where $G(x)$ is the incomplete depth between a preset reference plane and the bed plane, giving $Z_b(x)=G(0)-G(x)$.The initial water height and velocity are $g(x,0)=G(x)$ and $u(x,0)=0$. At the channel's inflow and outflow, we define $g(0,t)=20-4\sin\left[\pi\left(\frac{4t}{86400}+\frac{1}{2}\right)\right]$ and $u(L,t)=0$ respectively. The asymptotic analytical solution for the short tidal wave is given by '[6]':

$$g(x,t)=G(x)+4-4\sin\left[\pi\left(\frac{4t}{86400}+\frac{1}{2}\right)\right],\ u(x,t)=\frac{(x-14000)}{5400g(x,t)}\cos\left[\pi\left(\frac{4t}{86400}+\frac{1}{2}\right)\right] \tag{11}$$

The D2Q9 velocity model is used. The slip or non-slip boundary conditions are used at the solid walls. For the non-slip condition, the bounce-back plan is used and for slip conditions, a zero gradient of the distribution function perpendicular to the solid wall is employed and periodic boundary conditions are applied in the upper and lower walls. To achieve a lattice-independent solution, lattices of 15000×6000 and the lattice speed $e=200m/s$ and $\tau=0.6$ are also used. The results are given in 'figures 2-5'. This confirms the accuracy of the model for unsteady shallow-water flow problems. The present method can provide solution of equal accuracy as found in '[6]', where they used a complex upwind discretisation for the source term of the bed slope.

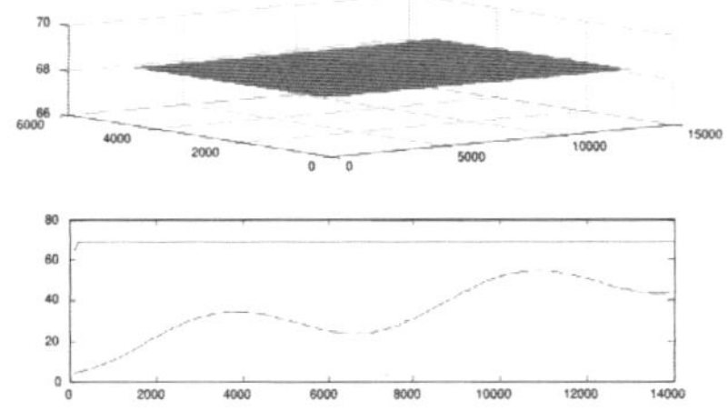

Figure 2 Numerical tidal free surface wave flow at time t=0.0

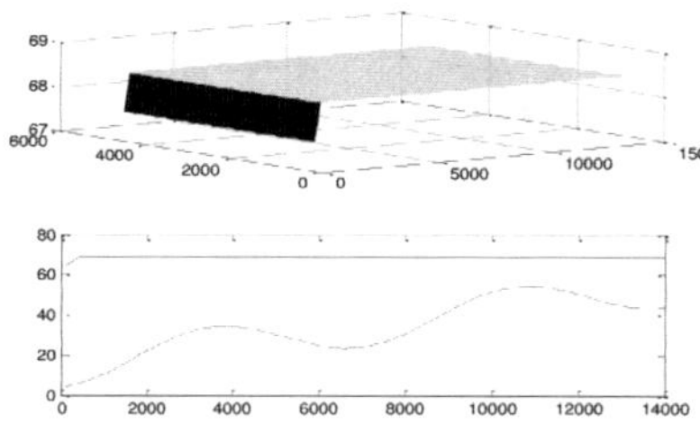

Figure 3 Numerical tidal free surface wave flow at time t =1.00.

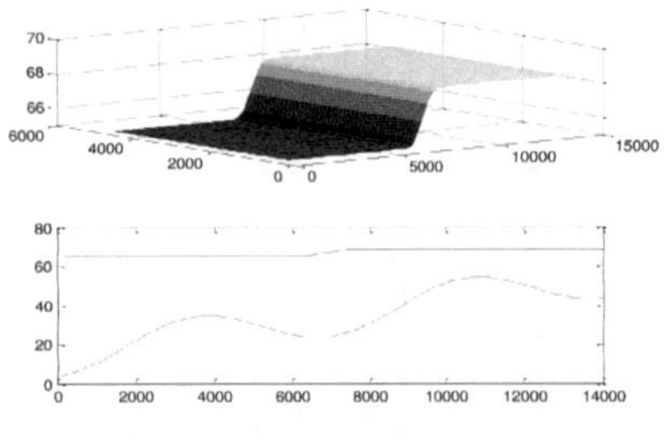

Figure 4 Numerical tidal free surface wave flow at time t=50.0.

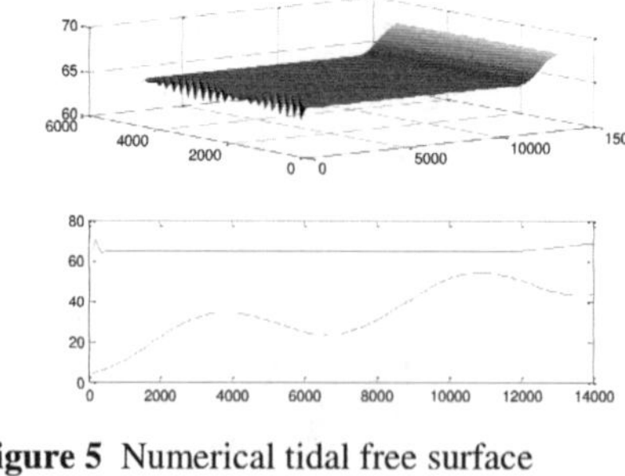

Figure 5 Numerical tidal free surface wave flow at time t = 100.00.

3.2 2D Tidal flow over an irregular bed in 3D plot

Let us consider a tidal flow that occurs over a bed that is known not to be regular as further test for the capability of the LBM for shallow-water equation. The bed is the same as that defined in 'Table 1'. For numerical computations, the D2Q9 velocity model is used with f^{eq} defined.The structure of the grid contains 150×60

lattice points with, the initial and boundary conditions are $g(x,0)=16-Z_b(x)$ and $u(x,0)=0$ and $g(0,t)=20-4\sin\left[\pi\left(\frac{4t}{86400}+\frac{1}{2}\right)\right]$ and $u(L,t)=0$, and thus the asymptotic analytical solution of the short tidal flow is

$$g(x,t)=20-Z_b(x)-4\sin\left[\pi\left(\frac{4t}{86,400}+\frac{1}{2}\right)\right],\ u(x,t)=\frac{(x-L)\pi}{5,400g(x,t)}\cos\left[\pi\left(\frac{4t}{86,400}+\frac{1}{2}\right)\right] \quad (12)$$

The centred scheme is employed for the force term. In order to contrast the numerical results and the asymptotic analytical solution, we select two results at $t=10,800s$ and $t=32,400s$, which relate to the half-risen tidal flow with maximum positive velocities and to the half-ebb tidal flow with maximum negative velocities and presented the results as shown in 'figures 6-8'. These figures shows that the numerical calculations and that obtained analytically are excellently in agreement. This supports the claim that the centred scheme is likewise correct and conservative for tidal flow that occurs over a bed that is known to be irregular. The results acquired with the basic and second order schemes are also carried out. Comparisons are done for the water surface and maximum positive velocities at $t=10,800s$. It is obvious from the figures that only the centred scheme yield accurate result.

Table1 Bed elevation z_b at point x for irregular bed.

x	0	50	100	150	250	300	350	400	425	435	450	475	500	505
z_b	0	0	2.5	5	5	3	5	5	7.5	8	9	9	9.1	9
x	530	550	565	575	600	650	700	750	800	820	900	950	1000	1500
z_b	9	6	5.5	5.5	5	4	3	3	2.3	2	1.2	0.4	0	0

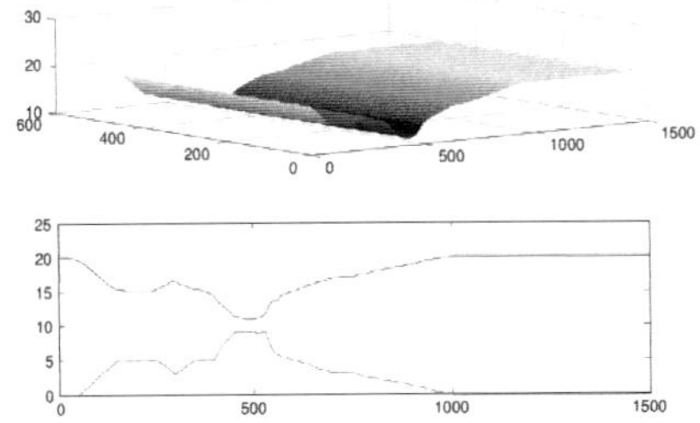

Figure 6 Numerical free surface for tidal flow over an irregular bed at time t = 1.0.

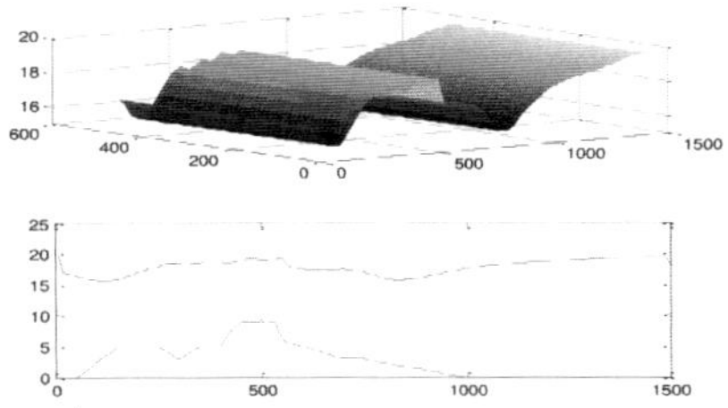

Figure 7 Numerical free surface for tidal over an irregular bed at time t = 30.0.

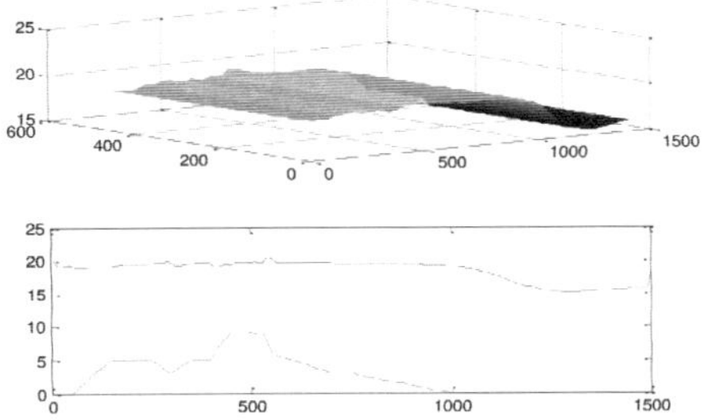

Figure 8 Numerical free surface for tidal flow over an irregular bed at time t = 70.00.

3.3 2D steady flow over an irregular bed in 3D plot

The bed landscape is described in 'tables 1-3'are shown in 'figures 9-12'.

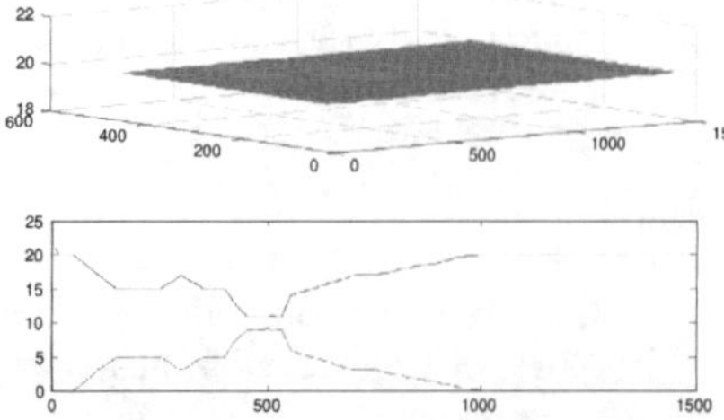

Figure 9 Numerical free surface for steady flow over an irregular bed at time t = 1.00

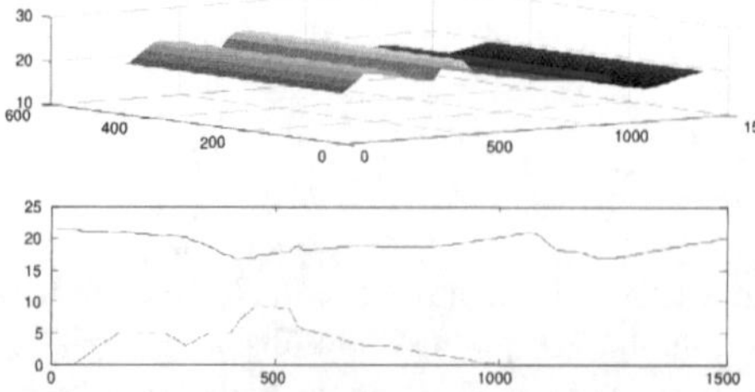

Figure 10 Numerical free surface for steady flow over an irregular bed at timet = 2.00

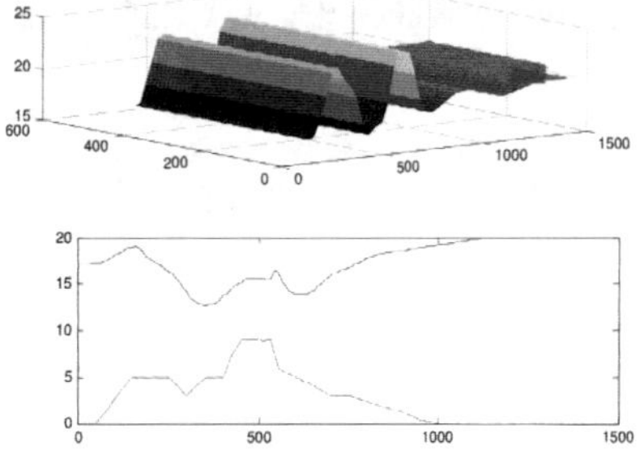

Figure 11 Numerical free surface for steady flow over an irregular bed at time t = 10.00.

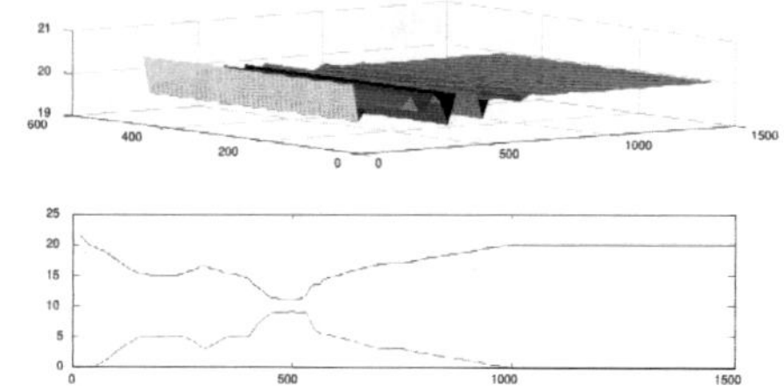

Figure 12 Numerical free surface for steady flow over an irregular bed at time t = 60.00

Table 2. Values of various parameters used for the wave Propagation over an oscillatory bottom test-case.

Initial wave number k	$1\,m^{-2}$
Gravity acceleration g	$1 \times ds\ ms^{-2}$
Initial wave amplitude b	$0.2\,m$
Undisturbed water depth d_0	$1\,m$
Bathymetry oscillation amplitude α	$0.001\,m$
Low bathymetry oscillation wavelength k_2	$2\,m^{-2}$
High bathymetry oscillation wavelength k_z	$6\,m^{-2}$
relaxation time τ	1
Real Channel Length rx	20

Table3. Values of various parameters used for different test-case.

Values of various parameters used	Tidal Flow over a Regular Bedtest-case.	Tidal Flow over an Irregular Bedtest-case	Steady Flow over a Bumptest-case	Steady Flow over an Irregular Bedtest-case
relaxation time τ	0.6	1.5	1.1	0.9
Real Channel Length rx	14000m	1500m	2m	1500m
dx	0.05m	0.05m	0.05m	0.05m
h_0	64.5m	20m	1.8m	20m
Gravity acceleration g	$9.81\times dx$	$9.81\times dx$	$9.81\times dx$	$9.81\times dx$
u_x	0	0	0.005	0.05
u_y	0	0	0	0

4 Conclusions

A new lattice Boltzmann model is suggested to solve the *2D* NSW wave equations. The efficiency and accuracy of the model are confirmed via thorough numerical simulation with lattice Boltzmann equation. It is noted that in order to attain better accuracy the LABSWE requires a relatively small time step Δt and the proper range is from 10^{-3} to 10^{-4}. The work would like to underline the importance of a robust runup algorithm development using the current model. This research should shift forward the accuracy and comprehension of a water wave runup onto complex shores. The results obtained reveal that NSW equation has sufficient prediction ability for maximum runup value. In conclusion, we have used three examples to test the LBM. It can be concluded that LBM performs well for such problems. The numerical results agree with the theory and hence, one can conclude that the stability structure is a good tool for designing the LBM. Furthermore, on the time-dependent problems on the unsteady problem, excellent and accurate results are obtained with no additional steps on the source terms or complicated upwind discretization of the gradient fluxes.

5 Acknowledgments

This work is supported by MOHE, Flagship Project Vote No. 01G40 under Research Management Center (RMC), Universiti Teknologi Malaysia, Johor Bahru, Malaysia.

REFERENCES

[1] Yaacob N, Aziz Z A and Norhafizah M S 2008 Modelling of Tsunami waves. *Matematika*. **24** 211-230.

[2] Yoon T H and Kang S K 2004 Finite Volume Model for Two-Dimensional Shallow Water Flows on Unstructured Grids. *Journal of Hydraulic Engineering*. **130** 678-688.

[3] Frandsen J B 2006 Investigations of wave runup using a LBGK modeling approach. *In CMWR XVI Proc. XVI Intl. Conf. Comp.Meth.in Water Resoures,* Copenhagen, Denmark.

[4] Th¨ommes G, Seaid M and Banda M K 2007 Lattice Boltzmann methods for shallow water flow applications. *Int. J. Numer. Meth.Fluids*. **55** 673–692.

[5] Zhou J G 2002 A lattice Boltzmann model for the shallow water equations. *Computer Methods in Applied Mechanics and Engineering*. **191** 3527-3539.

[6] Bermudez A and V´azquez M E 1994 Upwind methods for hyperbolic conservation laws with source Terms. *Comput. Fluids*. **23**, 1049.

IOP Conf. Series: Earth and Environmental Science **23** (2015) 012008 doi:10.1088/1755-1315/23/1/012008

Technological solution for vulnerable communities: Questioning the sustainability of Appropriate Technology

CPM Sianipar[1,2,*], K Dowaki[1], G Yudoko[2]

[1] Department of Industrial Administration (IA), Tokyo University of Science (TUS), 2641 Yamazaki, Noda-shi, Chiba-ken 278-8510, JAPAN

[2] School of Business and Management (SBM), Institut Teknologi Bandung (ITB), Jl. Ganeca 10, Bandung, West Java 40132, INDONESIA

*Email: morgana.sianipar@sbm-itb.ac.id; http://morganasianipar.com

Abstract. Vulnerability eradication has become an emerging concern in today's society following the increasing uncertainties in achieving societal resilience, particularly in vulnerable communities. Furthermore, incorporating technological solution, especially appropriate technology (AT), into such concern requires interdisciplinary understandings to achieve a holistic eradication based on the particularities of each community. This study aims to briefly reveal existing scholarly discourses and investigate potential gap(s) between previous researches. Literatures, particularly consisting meta-analysis on previous scholarly discussions, are surveyed. The findings reveal three progress among scientific discourses. The first one is the paradigm shift of developmental purposes from typical development to empowerment. Next, concerns in technology development indicate the parallel movement toward empowerment. Then, previous methodological developments, including approach in sustaining AT, indicate the needs to assess the future based on sustainability. Therefore, a new research is proposed to develop an assessment framework on AT for vulnerability eradication on the basis of empowerment paradigm, extended focuses in technology development, and extended coverage of future changes in dynamic matter. The framework needs to be developed based on the combination of positivist-deductive-qualitative research paradigms. This is intended to generalize the framework for being used in different cases, to build an applicative framework as an integral part of existing body of knowledge, and to develop an enriched and flexible construction of framework. Looking at existing researches, this brief study proposes insights to move scientific progress toward a more holistic vulnerability eradication using AT solution both in conceptual and practical levels.

1. Introduction: Technological changes for vulnerability eradication

The increasing uncertainties in today's complex world have shifted attentions of scholars and practitioners to vulnerability-related issues. While discourses on world's worst problems, *i.e.* poverty or environmental hazard, focus on the importance of distinguishing between symptoms and root problems in order to do decision making for taking strategic problem solving, vulnerability-related studies propose a more integrated understanding to put any societal attributes as having reciprocal influences each other, creating a holistic understanding in eradicating vulnerabilities. The term vulnerability itself has an interdisciplinary understanding [1-2] due to those reciprocal influences between fundamental factors in an observed societal group. Furthermore, the critical position of vulnerabilities in exposing a societal group to crises has made concerns on vulnerability eradication to target developing communities [3-4]. These kind of communities is recognized as the most critical type of communities which might fall into crises due to the instability of its fundamental factors of survivability. Vulnerabilities critically embedded in a particular community then makes the community to be stated as a vulnerable one.

On the other hand, implementing vulnerability eradication with an additional attention to the particularities of each vulnerable community requires an interdisciplinary solution. In such understanding, technology has emerged as a strong solution to eradicate vulnerabilities directly from interconnected characteristics of a technology to many facets of daily routines of local people [2-3,5]. Also, particularities in vulnerability eradication has triggered the further potential of Appropriate Technology (AT) to be a powerful technological solution. AT itself is an approach for providing a technological solution that is founded on the particularities of addressed problems in a specific area. Its strongest power is stated as technological appropriateness, by which the solution is developed based on local problem solving and matched to specific requirements on field since the earliest stage of technological development. In facts, characteristics of every vulnerable community as a fragile societal entity indicate the need of a technological solution which has strong technological appropriateness to precisely address its vulnerabilities [3]. Furthermore, as vulnerability eradication is considered as a continuous process, there is also a need to apply technological changes. The solution is therefore stated to incorporate technological changes on AT to conduct a sustainable vulnerability eradication. Based on such concerns, recent developments on both technological changes and AT have proposed an important progress toward vulnerability eradication [6]. In technological changes, recent discourses have begun to propose a more sustainable change based on bottom-up approach. Researchers agreed that technology developments need to incorporate global concerns, yet the actions must be local in order to seamlessly diffuse those concerns into local activities. Those understandings are parallel with the development of thoughts on AT. When concerns should be global yet actions are local [7-8], AT is the right way to do those combination. Recent developments of AT indicate that its concept has been widely recognized in putting technological appropriateness as an ultimate characteristic of any technologies. The concept of AT is hence diffused also into any effort to produce technological solutions for vulnerability eradication.

However, those scientific progress – the parallel developments of conceptual thoughts, field researches, and technological solutions that contribute to the evolution of vulnerability eradication over time – are rather discrete than unified. Despite their parallel developments, they require new researches to move the progress toward both unified conceptual understandings and practical actions. Vulnerability eradication using AT as the solution, therefore, is first waiting the answers for these following questions:

RQ1 How scientific progress relate each other? What kind of further researches required?

RQ2 How researchers are advised for conducting the kind of researches revealed from the RQ1?

2. Research gap: Dynamic assessment below the radar

The scholarly discussions on the paradigm shift in the study of technology, particularly technological changes, for vulnerability eradication in vulnerable communities [9-10] have indicated that there is an ongoing progress in scientific communities to put empowerment as the ultimate intention of developmental works [9,11-12] (Figure 1). In such discourse, the intention is to put local people as the subject of development, meaning that they become the main conductor of cyclical developmental progress in their own future. The vulnerability eradication is hence taken as a sustainable process which covers future decisions related to changes on vulnerabilities as the root problem and changes on applied solutions taken in the eradication. Furthermore, when technology, particularly AT, is taken as the solution in vulnerability eradication, the shift of paradigms also affects the focus of technology development [6,9] (Figure 1). In the typical development paradigm, technology development is only focused on two matters: how a technology is designed by foreign partner(s) and is introduced to general vulnerable communities. Indeed, technology is seen as merely a product of design process behind-a-closed-door that would be taken as a solution for vulnerability eradication by local people in their local activities. On the other hand, the shift to empowerment paradigm requires an extended focus of technology development [13] to cover more applicative activities such as the usage, local repair, re-usage, and local disposal of an AT. Those four activities are the minimum requirements of observation in understanding the sustainability of a technological solution. Besides, some other particular process, *i.e.* material degradation, trial & error, *etc.*, are possible to be incorporated if necessary. By covering a wider observation, analysis on the sustainable technological appropriateness and predicted changes on vulnerabilities could be conducted through an explicative and more holistic investigation.

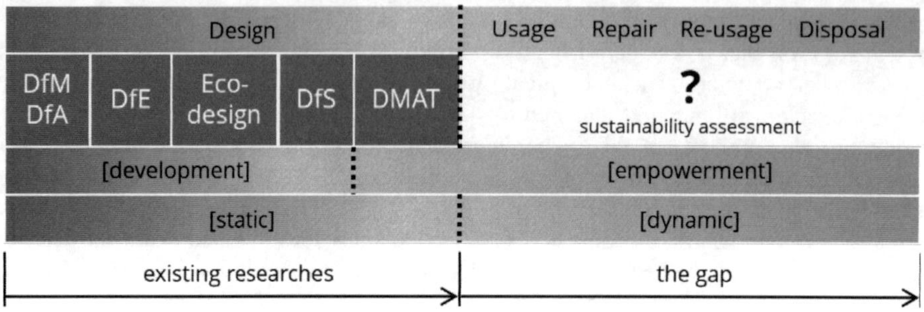

Figure 1. The research gap: Dynamic assessment.

While the shift of paradigms is widely discussed in terms of purely conceptual understandings, scientific progress on methodological development [6] is also increasingly interesting for scholarly communities. However, the shift is not devoted to completely replace previous methodologies with a new one. In other words, applied paradigms in different methodologies are rather incremental than absolute, meaning that many newly developed methodologies are intentionally purposed to cover more specific area to refine the coverages, qualities, and characteristics of previous ones. Based on the spirit to refine the coverage of previous "Design for X" methodologies, Eco-design and DfS [14-15] have initiated a shift towards empowerment-based ones. Despite the critical focus of Eco-design and DfS on environmental issues, there are big parts of their approaches still stand on typical development paradigm [9]. After that, DMAT emerges to clearly bring technology development into a more empowerment-based one [16]. The purposeful methodology is stated as an important breakthrough to put local people as the subject of development. Therefore, the DMAT could be distinguished to previous methodologies as it stands on empowerment paradigm (Figure 1). However, either DMAT or DfS and previous ones are strongly focused on design process [6]. In spite of the improved coverage of DfS and DMAT to consider future changes of technology usage and affected prosperity of its users, their focuses are intentionally directed on the design process. Other applicative process are hence posited as additional considerations, yet the investigation takes static calculation on present situations to predict the future.

On the other hand, investigating extended focuses of technology development for vulnerability eradication, including the shift from typical development to empowerment, also requires an extended analysis that covers many possible changes in the future. While present situation changes in the future, addressed vulnerabilities and applied solutions would also change to be in conformity with required actions in each circumstance [2-3,17-20]. Indeed, those changes may happen as several dynamic possibilities depend on some possible scenarios. In fact, those possibilities are parallel to the understanding of shift toward empowerment-based analysis. Scientific progress on pure conceptual discourses and refined focuses on methodological development indicate that the development of a technological solution for vulnerability eradication in a particular vulnerable community ultimately requires dynamic assessment on some possible scenarios which incorporate alternative solutions possibly happen or taken in the future (Figure 1). In such understanding, the sustainability of both technological solution, particularly AT, and vulnerability eradication determines the resilience of a particular community. In other words, the sustainability of solution and process produces the survivability of targeted societal group. Looking at the existing scientific developments (Figure 1), there is a lack of research on the dynamic analysis to assess the sustainability of AT and affected prosperities of local people. Therefore, research is required to develop a new assessment framework/tool for conducting dynamic investigation on alternative technological solutions, particularly AT, as a means to understand the dynamic behavior of the solutions in a sustainable manner. The research needs to cover both required focuses based on both development and empowerment with more focus on the later paradigm. Then, the product of the research should be in line with the ongoing scientific progress, meaning that the new dynamic assessment would become an integral part of the existing body of knowledge of technological solution for vulnerability eradication in vulnerable communities.

3. Research scopes and positioning

Based on the research gap, including required research to fill in the gap, there are several scopes which need to be taken as the guidance of analysis. An adequate research scope means that the required research would consist of enriched contents without ignoring the appropriateness of covered boundary. Looking at previous discussions, there are four groups of optional coverages (Figure 2). The incorporated coverages are then selected among options in each of those groups to indicate the research positioning among available combination of coverages. The first group is the developmental paradigm, consisting aforementioned typical development and empowerment ones. As previously discussed, the selected paradigm is empowerment. In spite of possible coverage on some characteristics of typical development, empowerment paradigm is intentionally selected as the basis of the new dynamic assessment framework to be parallel with recent scientific progress. The purpose is also affected by the concern of putting local people as the subject of development, creating a more survivable societal group with a good resilience.

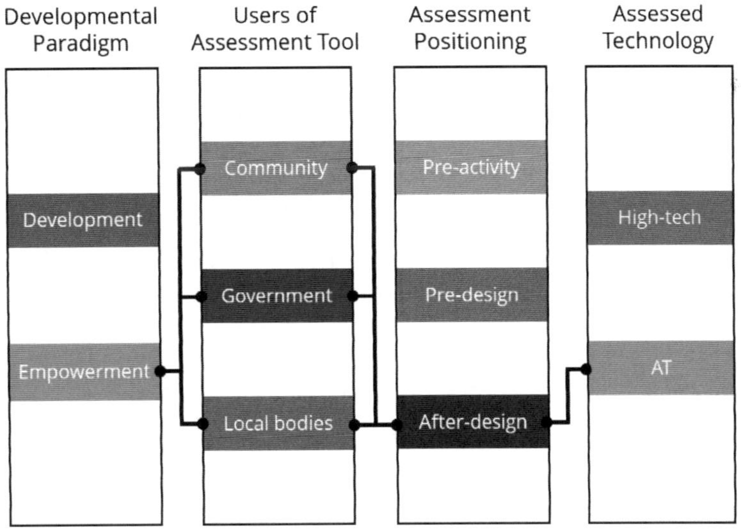

Figure 2. Research scope & positioning.

Next, the second group is three possible users of the new assessment framework/tool, namely Community, Government, and Local bodies. Considering the usefulness of a dynamic assessment, those three options are selected. For Community, a dynamic assessment means their capability to understand the effects of any proposed ATs on their future prosperity. For Government, assessing alternative technologies through a dynamic analysis is critical to ensure a continuous vulnerability eradication in their jurisdiction. Then, local bodies such as NGOs or association could take the advantage to offer a more holistic assistance for local people. After that, regarding the positioning of assessment process itself (Figure 2), the option pre-activity means the assessment is taken before the decision of using technology as the solution for eradicating vulnerability. The second option, pre-design, refers to the decision taken to choose existing alternative technologies as the basis of design. Then after design is taken in deciding which AT would be applied among some alternatives. By considering the research gap and the intention of dynamic analysis, assessment is only posited to cover after-design, meaning that dynamic assessment is taken on some alternative technologies that have been designed in order to choose the best design for being applied further. Moreover, the last group indicates the type of technology covered in the research (Figure 2). The framework/tool is intentionally purposed to assess AT as the technological solution, hence high-tech is not covered by the research.

4. Research paradigms

Besides the scope and positioning, a research requires a set of beliefs and assumptions taken as the base reference in conducting step-by-step research activities [21-22]. To distinguish applied paradigm(s) taken in current research (Figure 3) with applicable paradigm(s) of research product (Figure 1), existing research paradigms are categorized into three kinds, namely paradigm itself, paradigmatic approach, and paradigmatic method [23-24]. Roughly-speaking, research paradigm indicates the standpoint of current research in developing the dynamic assessment framework/tool (research product), while paradigm of research product refers to the standpoint taken as the basis of incorporated understandings in the framework [9]. For each paradigm category, there are two options to simplify many available options throughout scholarly literatures. The understanding hence requires non-dichotomy which refers to approximate meanings of each option. In other words, both options are legitimate and valuable in conducting any research. In the first category (paradigm), two options are available, namely constructivism and positivism [21,25-26]. Considering the purpose of assessment development to produce distinguishable scientific contribution to existing body of knowledge, positivism is selected [27]. Thus scientific discovery could be maintained on a single cycle of research with a focus on the top-down side of research wheel [24]. Also, the development of assessment framework is posited in a controlled reality to avoid bias perception if constructivism is taken [24,28]. In the paradigmatic approach category, deductive approach follows positivism to strengthen the top-down development of assessment framework [24]. However, this needs to be distinguished with the application of assessment framework as a more bottom-up usage based on empowerment paradigm. In terms of deductive approach, assessment framework is constructed based on analysis on scholarly literatures in order to establish an applicative ones based on existing discourses [25]. The inductive paradigmatic approach is not selected to avoid an infinite cyclical development process of the framework, including some possible particularities as the strengths of inductive approach which may prevent the framework to be a generalized model [24-25]. Then, the research is intended to be a more qualitative one, meaning that analysis on existing literatures, including any types of analysis taken as the process guidance, is conducted based on qualitative method. The quantitative method is not selected to avoid tricky analysis that is possible to bring a prolonged development process within research timeframe [24-25]. However, dynamic assessment as the research product is intended to be a quantitative framework.

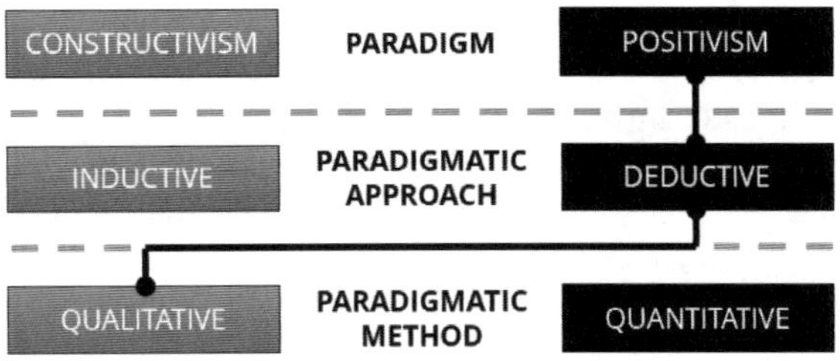

Figure 3. Paradigms, paradigmatic approach, & paradigmatic method.

5. Conclusion

Vulnerability eradication using AT as an interdisciplinary technological solution requires a unified understanding of those idea. Among related concepts, there is a parallel scientific progress towards a holistic approach in conducting the eradication. The main conjunction is the shift from typical development paradigm to empowerment. It is the driving force behind other developments in scholarly discourses. In short, there is a need to develop a new dynamic assessment framework for deciding the most appropriate technology for vulnerability eradication. The dynamic assessment would extend the focuses of both technology and methodological developments to cover more empowerment-based understandings. Furthermore, a combination of positivist-deductive-qualitative research paradigms in the development of the proposed framework is advised in order to: develop an enriched yet flexible construct of the framework; produce a generalized framework that is possible to be used in different cases; build the framework as an integral part of existing scientific progress. Moreover, every explanation in this brief study is critical to be posited as the basis of any future researches related to the methodological development for vulnerability eradication in order to push scientific progress towards an incremental and integrated development of both conceptual understandings and practical actions.

References

[1] Cutter S L, Boruff B J and Shirley W L 2003 *Soc. Sci. Quart.* **84** (2) 242-261
[2] Turner B L, Matson P A, McCarthy J J, Corell R W, Christensen L, Eckley N, Kasperson J X, Luers A, Martello M L, Polsky C, Pulsipher A, Schiller A, Selin H and Tyler N 2003 *P. Natl Acad. Sci. USA.* **100** (14) 8080-8085
[3] Sianipar C P M, Dowaki K and Yudoko G 2014 *IOP Conf. Ser. Mater. Sci. Eng.* **58** 012022
[4] Robards M and Alessa L 2004 *Arctic.* **57** (4) 415-427
[5] Garniati L, Owen A, Kruijsen J, Ishadamy Y and Wibisono I 2013 *Sustain. Cities. Soc.* **12** 9-15
[6] Sianipar C P M, Yudoko G, Dowaki K and Adhiutama A 2014 *J. Sustain. Sci. Manage.* **9** (1) 1-17
[7] Few R 2003 *Prog. Dev. Stud.* **3** 43-58
[8] Leichenko R M and O'Brien K L 2002 *Mitig. Adapt. Strat. Global. Chang.* **7** 1-18
[9] Sianipar C P M, Yudoko G and Adhiutama A 2014 *Adv. Sci. Lett.* **20** (in press)
[10] Heinen J T 1994 *Int. J. Sustain. Dev. World Ecol.* **1** 22-33
[11] Narayan D 2005 *Measuring empowerment: Cross-disciplinary perspectives* (Washington D.C., US: World Bank)
[12] Alsop R, Bertelsen M F and Holland J 2006 *Empowerment in practice: From analysis to implementation* (Washington D.C., US: The World Bank)
[13] Sianipar C P M, Dowaki K, Yudoko G and Adhiutama A 2013 *Eur. J. Sustain. Dev.* **2** (4) 1-18
[14] Fiksel J 1996 *Design for Environment* (New York, US: McGraw-Hill)
[15] Clark G, Kosoris J, Hong L N and Crul M 2009 *Sustainability.* **1** (1) 409-424
[16] Sianipar C P M, Yudoko G, Dowaki K and Adhiutama A 2013 *Sustainability.* **5** (8) 3382-3425
[17] Turner B L, Kasperson R E, Matson P A, McCarthy J J, Corell R W, Christensen L, Eckley N, Kasperson J X, Luers A, Martello M L, Polsky C, Pulsipher A and Schiller A 2003 *P. Natl*

Acad. Sci. USA. **100** (14) 8074-8079
[18] Richmond B 1993 *Syst. Dyn. Rev.* **9** (2) 113-133
[19] Bagheri A and Hjorth P 2007 *Sustain. Dev.* **15** (2) 83-96
[20] Hjorth P and Bagheri A 2006 *Futures.* **38** (1) 74-92
[21] Guba E and Lincoln Y 2005 Paradigmatic controversies, contradictions, and emerging influences *The Sage Handbook of Qualitative Research (3rd edition)* eds N K Denzin and Y S Lincoln (Thousand Oaks, CA: SAGE Publications) pp 191-215
[22] Morgan G 1983 Research strategies: Modes of engagement *Beyond method* ed G Morgan (Beverly Hills, CA: SAGE Publications)
[23] Haase J E and Myers S T 1988 *Western. J. Nurs. Res.* **10** 128-137
[24] Neuman W L 2006 *Social research methods: Qualitative and quantitative approaches* (Boston, US: Pearson Education)
[25] Jaccard J and Jacoby J 2010 *Theory construction and model-building skills* (New York, US: The Guilford Press)
[26] Krauss S E 2005 *Qual. Rep.* **10** (4) 758-770
[27] Schrag F 1992 *Educ. Res.* **21** (5) 5-8
[28] Mills J, Bonner A and Francis K 2006 *Int. J. Qual. Method.* **5** (1) 1-10

Enhancement in resistivity resolution based on the data sets amalgamation technique at Bukit Bunuh, Perak, Malaysia

Andy Anderson Bery*[1], Rosli Saad, I.N.E. Hidayah, I.N. Azwin
[1]Geophysics Section, School of Physics, Universiti Sains Malaysia, 11800 Penang, Malaysia.

Mokhtar Saidin[2]
[2]Centre for Global Archaeological Research, Universiti Sains Malaysia, 11800 Penang, Malaysia.

E-mail: andersonbery@yahoo.com.my

Abstract. In this paper, we have carried out a study with the main objective to enhance the resolution of the electrical resistivity inversion model by introducing the data sets amalgamation technique to be used in the data processing stage. Based on the model resistivity with topography results, the data sets amalgamation technique for pole-dipole and wenner-schlumberger arrays are successful in identifying the boundary or interface of the overburden and weathered granite. Although the electrical resistivity method is well known, the proper selection of an array and appropriate inversion parameters setting such as damping factors are important in order to achieve the study objective and to image the target at the Earth's subsurface characterizations.

1. Introduction

In the past few years, significant effort has been done in developing in-field data acquisition technique in electrical resistivity method. This method work by measuring the potentials between two electrodes while injecting current into the ground at another electrode pair. The enhancement of model resolution in resistivity method is challenging in geophysical imaging in order to characterize the Earth's subsurface. In addition, the identification of the Earth's subsurface structures which is invisible materials or structures undersurface of the Earth such as bedrock and alluvium thickness. This is based on amplitude and geometry indicators obtained via this method, requires a good understanding about the electrical resistivity arrays and appropriate parameter settings selection.

The electrical resistivity method has been used worldwide in engineering and environmental study. This geophysical method has been used for a long time in the investigation of the Earth's subsurface. The electrical resistivity method is usually used in bedrock determination, overburden materials and subsurface structure such as cavities, boulders, faults and voids [1]. In environment study, there are many physical parameters can take into consideration before making an important decision in civil construction. These physical parameters play an important role in indicating their characterization which is due to time and condition changes [2]. The in-situ behaviour of soils is complex because it is

[1] Geophysics Section, School of Physics, Universiti Sains Malaysia, 11800 Penang, Malaysia.
[2] Centre for Global Archaeological Research (CGAR), Universiti Sains Malaysia, 11800 Penang, Malaysia.

heavily dependent upon numerous factors. To acquire appropriate understanding, it is necessary to analyze them not only through geophysics and geotechnical engineering methods, but also through other associated disciplines like geology, geomorphology, climatology and other earth and atmosphere related sciences [3]. Thus, the correlation of these technical methods of investigation of Earth subsurface is required by geoscientists and engineers. As no single geophysical method yields the optimal information about the Earth subsurface, it is better to combine the geophysical method such resistivity tomography, or seismic refraction tomography with geotechnical works which is in-situ measurements. The main objective of this study is to enhance the resolution in the inversion model in electrical resistivity method. The development of the technique are discussed in the next section.

2. Geological setting

Bukit Bunuh is located in the Malaysian most famous prehistoric valley, Lenggong Valley, Perak, Malaysia. Among the most notable of the excavations of Paleolithic period is located at the Paleolithic sites at Kota Tampan, Lawin, Bukit Jawa and Temelong [4]. These prehistoric archaeological sites classified as open sites, with the materials of quartzite and quartz. Bukit Bunuh was found at the end of 2001, which also an in-situ undisturbed open sites with the tool making workshop evidences [5]. At Bukit Bunuh, the quartz blocks can be classified as large to small fragments which forming suevite rocks.

3. Materials and methods

The electrical resistivity survey is a surveying technique for imaging the Earth's subsurface characterization including complex geology structure. In this study, we carried out an electrical resistivity survey using multi-electrode system. In this system, we are using 41 electrode number. The current and potential electrodes are changed automatically by Electrode-Selector (ES10-64), which is based on selected array for the resistivity survey. The minimum electrode spacing used in this study is 5 m spacing and the total length of straight survey line is 200 m. There are three different resistivity arrays used in this study, which is pole-dipole, dipole-dipole and wenner-schlumberger arrays. Each of this resistivity arrays has their own advantage and limitations in imaging the subsurface. Thus, the inversion model obtained using RES2DINV software is correlated with the borehole record which acts as geological reference. This will help in the interpretation work of the inversion resistivity models obtained to be more reliable and effective. In this study, we are studying three possible data sets amalgamation for three selected different types of resistivity arrays. Figure 1 shows the schematic diagram for possible combination of all three selected resistivity arrays. Simply codes were created by these possible data sets amalgamation technique are given as (P-D) for pole-dipole array, (D-D) for dipole-dipole array and lastly (W-S) for wenner-schlumberger array. The code combination used are (a) pole-dipole and dipole-dipole arrays (P-D and D-D), (b) dipole-dipole and wenner-schlumberger arrays (D-D and W-S) and (c) pole-dipole and wenner-schlumberger arrays (P-D and W-S).

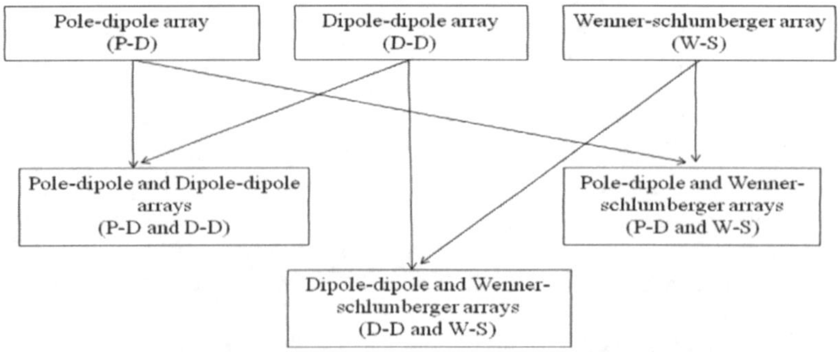

Figure 1: Schematic diagram for possible data sets amalgamation technique for three different resistivity arrays used in this study.

Dipole-dipole array was sensitive to horizontal changes however, lower signal strength compared with pole-dipole. Pole-dipole array has good depth of penetration among all arrays, but signal strength is lower compared to wenner-schlumberger. Signal strength for wenner-schlumberger array is twice than pole-dipole, however horizontal coverage is narrower than dipole-dipole. Wenner-schlumberger array is the amalgamation of wenner alpha array and schlumberger array. This is proven by data coverage is similar between wenner alpha and schlumberger arrays amalgamation with wenner-schlumberger array. Thus, this array can be considered as co-linear and asymmetric array. In this paper, we are also presenting the real datum points to show the development of the data sets amalgamation technique used in this study. Thus, we will show that this technique is able to give significant advantages compared using one single array in engineering and environmental investigation. Figure 2 shows the datum points for pole-dipole, dipole-dipole and wenner-schlumberger arrays used in this study. The maximum number of datum points for pole-dipole array is 1387, dipole-dipole array is 356 and wenner-schlumberger array is 665.

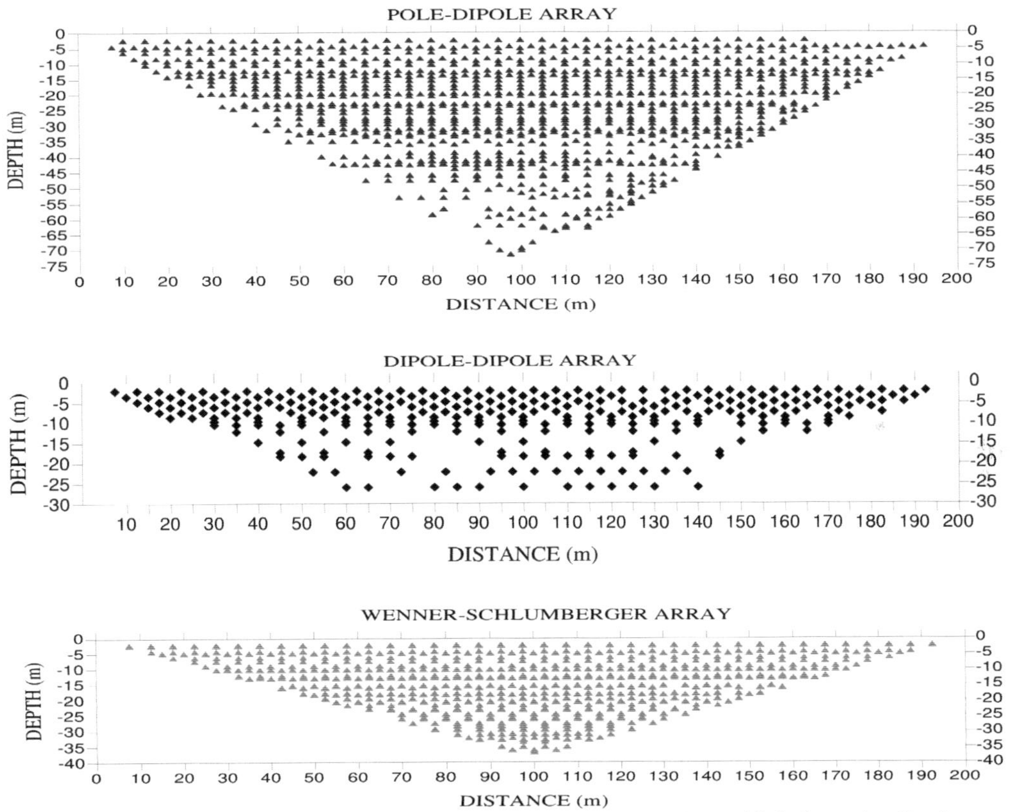

Figure 2: Three different resistivity arrays used in this study, which is pole-dipole array (above), dipole-dipole array (middle) and wenner-schlumberger array (bottom).

Meanwhile, Figure 3 shows the data sets amalgamation for all three different resistivity arrays used for inversion processing stage. By this technique, we have identified that there is an improvement in the maximum number of datum points used for 2-D inversion processing which is given by datum points summation of two different resistivity arrays. The maximum number of datum points identified in this study was (P-D and D-D) is 1743, (D-D and W-S) is 1021 and lastly (P-D and W-S) is 2052.

With this technique, it shows that the data levels of these two different arrays in the inversion models are improved better compared to one single array. Another term for this condition, we can consider that the enhancement in vertical resolution has occurred in inversion models. Moreover, the datum points in the horizontal section which is governed by electrode spacing is filled with other data sets of array. Thus, the horizontal resolution of the inversion model is enhanced using this technique.

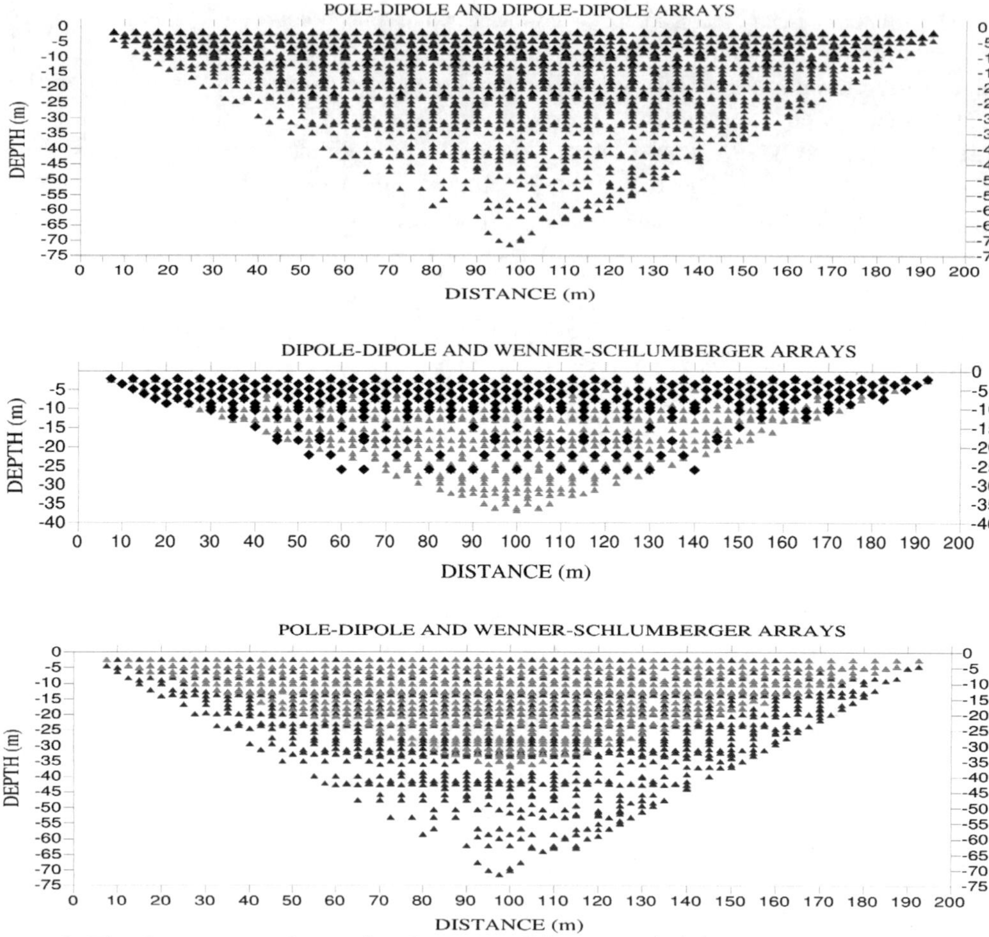

Figure 3: The data sets amalgamation from two different resistivity arrays used in this study. The data set used in the inversion models were (P-D and D-D) (above), (D-D and W-S)(middle) and (P-D and W-S)(bottom).

4. Results and discussions

In this paper, we present the results of the inversion resistivity model where the electrical resistivity survey was conducted in Bukit Bunuh, Perak, Malaysia. The electrical resistivity survey was conducted using appropriate survey setting which is same for all selected arrays. Figure 4 shows the resistivity model result using the data sets amalgamation technique for dipole-dipole and wenner-schlumberger arrays. Figure 5 shows the resistivity model result using the data sets amalgamation technique for pole-dipole and dipole-dipole arrays. Meanwhile, Figure 6 shows the resistivity model result using data sets amalgamation technique for pole-dipole and wenner-schlumberger arrays. All the resistivity model results are correlated with the in-line borehole record (BH-6). This borehole record is used in electrical resistivity model interpretation because this borehole record assists the interpreter to identify the Earth's subsurface lithology at one point only. However, the borehole record helps in the interpretation where this borehole record can acts as geological reference. In Figure 4, 5 and 6, the electrical resistivity model results were successful in imaging Bukit Bunuh subsurface. The overburden of the study site is classified as sandy silt and sandy materials. These two materials made up the overburden with thickness range from 10 m to 23 m. Figure 6 shows significant and reliable results where the boundary (black dash line) of the overburden and weathered granite using data sets amalgamation for pole-dipole (P-D) and wenner-schlumberger (W-S) arrays. However, this boundary is not significant and unclear for others as shown in Figure 4 and Figure 5.

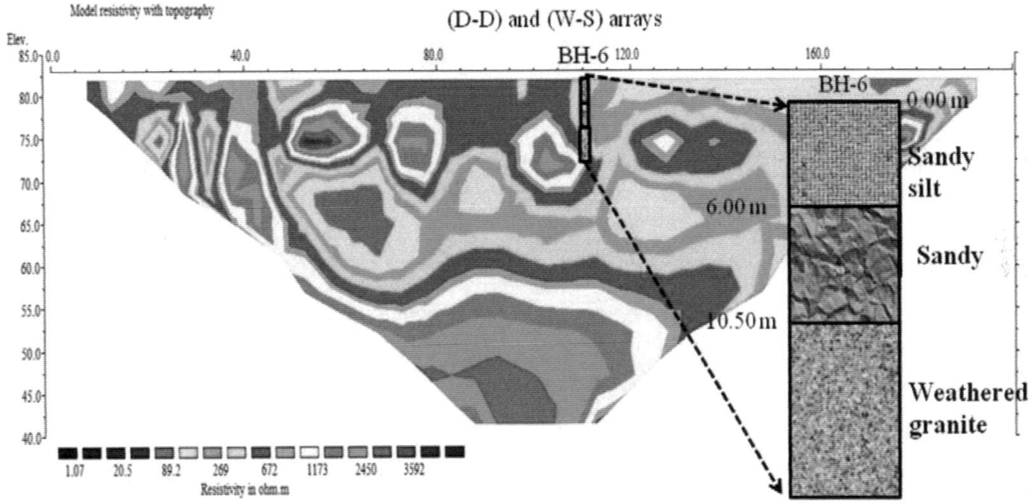

Figure 4: The model resistivity result for Bukit Bunuh subsurface characterization using (D-D) and (W-S) arrays along with geological reference (BH-6).

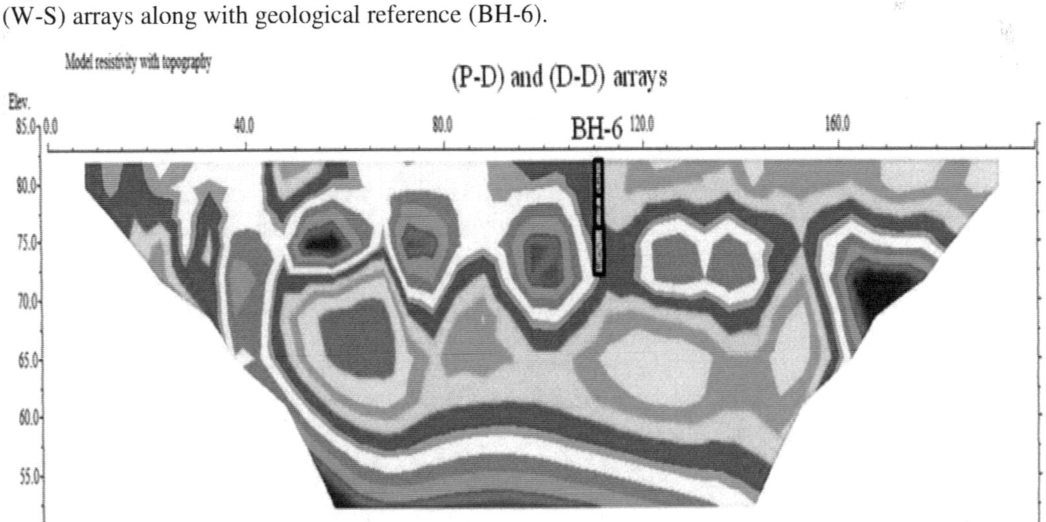

Figure 5: The model resistivity result for Bukit Bunuh area subsurface characterization using (P-D) and (D-D) arrays along with geological reference (BH-6).

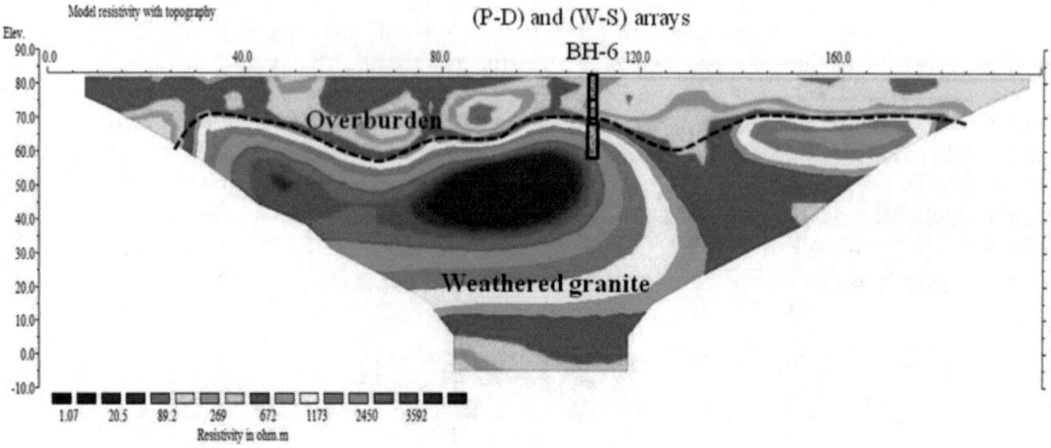

Figure 6: The model resistivity result for Bukit Bunuh subsurface characterization using (P-D) and (W-S) arrays along with geological reference (BH-6).

5. Conclusion

This electrical resistivity tomography method is a good geophysical tool for imaging the Earth's subsurface structure such as bedrock layer. In this study, we are successful achieved our objective to enhance the resolution in the resistivity model results using the data sets amalgamation technique. Although the time taken is twice in data acquisition works, it shows that this technique is applicable and has its technical merit in imaging the Earth's subsurface. Moreover, this study showed that this electrical resistivity resolution enhancement via this technique has covered both horizontal and vertical resolution in resistivity pseudo-section. To fully exploit the data, it is important for proper selection of an array and appropriate inversion parameters setting such as damping factors are used in this study.

6. References

[1] A. A. Bery, and R. Saad, "High resolution time-domain induced polarization tomography with merging data levels by two different optimized arrays for slope monitoring study", *Electronic Journal of Geotechnical Engineering*, vol. 18, bund. Z, pp. 5921-5928, 2013.

[2] A. A. Bery, and R. Saad, "High resolution time-lapse resistivity tomography wit merging data levels by two different optimized resistivity arrays for slope monitoring study", *Electronic Journal of Geotechnical Engineering*, vol. 19, bund. C, pp. 503-509, 2014.

[3] A. A. Bery, "Synthetic modelling application in high resolution seismic refraction tomography", *World Applied Sciences Journal*, vol. 28, issue 5, pp. 625-628, 2013.

[4] M. N. Nawawi, S. Mokhtar, I. A. Jaffrie, IA, S. Sarmiza, and A. Norazura, "Geophysical application in mapping Paleolithic workshop site in Bukit Bunuh, Perak, Malaysia", *Proceedings of The 7th SEGJ International Symposium-Imaging Technology*, pp. 427-430, 2004.

[5] M. Zuraina, "*Archaeology in Malaysia*", Universiti Sains Malaysia, Penang, 2003.

Acknowledgments

Andy A. Bery would like to thank geophysics postgraduate student and laboratory assistant, Mr. Yaakub Othman for their assistance in data acquisition works. The author also wants to thank the Centre for Global Archaeological Research (CGAR) Malaysia, Universiti Sains Malaysia for sponsoring drilling works at Bukit Bunuh site. Lastly, the author would like to give appreciation to Jeff Steven and Eva Diana for their encouragement.

IOP Conf. Series: Earth and Environmental Science **23** (2015) 012010 doi:10.1088/1755-1315/23/1/012010

Theoretical Explanations of Environmental Motivations and Expectations of Clients on Green Building Demand and Investment

Onuoha Iheanyichukwu Joachim[1], Norhaya Kamarudin[1], Godwin Uche Aliagha[1] and Kalu Joseph Ufere[1]

[1]Department of Real Estate Faculty of Geoinformation and Real Estate Universiti Teknologi Malaysia, 81310 Skudai, Johor, Malaysia

Email: onuoha4lord@gmail.com

Abstract

In building industry, green demand and investment creates a positive footprint on the environment. However, these environmental opportunities have not been adequately harnessed and explored by Clients of green building apparently because of poor understanding of the motivating drivers and benefits accruing from green building demand and investment. The decision to demand for or invest in green building is influenced by certain environmental motivating drivers and expectations which have not been fully examined by researchers and not well understood by stakeholders. Based on the Theory of Planned Behaviour (TPB) and Theory of Value Belief Norm (VBN) explanations, this study focused on the Clients, purchasers and users' motivations and intentions to go for green building. Based on the reviewed theories, we hypothesized that environmental motivations and expectations for green building demand and investment are embedded in the environmental quest for protection of eco-system and bio-diversity, improvement of water and air quality, reduction of solid waste, conservation of natural resources, reduction of societal costs of landfill creation and maintenance, minimization of site impact and reduction emission to air and enhanced energy efficiency. However, the predictive validity of these propositions depends on the client's beliefs, values, social pressure, and perceived behavioural control.

Keywords: Green building, Green building demand and investment, Environmental motivations and expectations, Theory of planned behaviour, Theory of value belief norm, Environmental sustainability.

1. Introduction

Green demand and investment is a term that comprises any property investment that gives the opportunity for investment that is environmentally responsive and takes into account of its effect on the environment. [1] referred green investment as an investment necessary to reduce greenhouse gas and air pollutant emission without significantly reducing the production and consumption of non-energy goods. For green investment in building to be meaningful, it should incorporate sustainable consideration and meet the demands of environmentally conscious buyer based on known driving benefits. The decision to demand and invest in green buildings are influenced by certain motivating drivers and expectations rooted on environmental behaviour and concern.

Environmental behaviour has been described to be a socially-conscious behaviour which is based on social responsibility and involves individual and social aims that one wants to achieve by behaving in a particular way [2, 3]. This suggests that for a person to accomplish what he intends to achieve, his or her actions are influenced by intentions and motivations with hope of expectations. Therefore to explain specific environmental consumer behaviour and concern, theories of Planned Behaviour (TPB) and Value Belief Norms (VBN) are considered. The unique attribute of both theories is behavioural intentions and motivations, values and norms which are recognized as the building block in predicting actual behaviour. Behavioural intentions have been viewed as an indication of individual's readiness to carry out a given behaviour [3]. In their study, [4] found that behaviour changes could come about through one's personal commitment and motivations. Behavioural motivation could therefore be seen as an inducement by internal or external factors to stimulate desire and energy in individuals to be continually interested and committed to perform a given behaviour. It is a psychological process that initiates, guides and maintains goal-oriented behaviours. Behavioural motivations involves emotional, social and cognitive forces that activate behaviour and which could be described as "Whys" of behaviour - the needs or wants that drive behaviour and explain what one does.

The Theory of Planned Behaviour (TPB) which is an extension of theory of Reasoned Action (TRA) differentiates two determinants that predict behavioural intentions and motivations. They include; personal attitudes towards behaviour and subjective norms which is an element of VBN. The first could be positive or negative assessment of a particular behaviour which is formed by people behavioural beliefs while the second reflects person's perception of social pressure regarding the performance of behaviour. [5] observed that this subjective norm is a function of individual's belief that referent others coupled with the individual's motivation to comply with these referent others. However this assertion starts from values which forms the basis for belief, which in turn underline norms which dictate behaviour as posited by [6] in the Value-belief norm theory(VBN).

The application of the two behavioural theories could induce pro-environmental habits, attitudes, and knowledge and arouse expectations and beneficial factors that could motivate clients and prospective users to go for green building or indulge in green practices. It could also guide their environmental actions and activities in prospect of future benefits. This study therefore explores those environmental motivating drivers and benefits, and grounds of expectations that could inform the understanding of the motivating factors of green building through the Theory of Planned Behaviour (TPB) and the Value-Belief Norm Theory (VBN).

2. Literature Review

2.1 Environmental Behaviour, Attitudes and Knowledge

Several studies on environmental behaviour have been carried out to define environmental behaviour but oftentimes environmental behaviour has been assumed to be an undifferentiated class of behaviour [7] .This simply means that various types of environmental behaviours are dependent on similar factors, which is not usually the case. Studies indicate that many psychological studies cantered on behaviours that are not very interesting from environmental perspectives.

To [7]; [8] and [9] the concentration is on individual behaviour that has only little effect on environmental qualities. Such behaviours in the views of [10] include refusing plastic bags in stores or purchasing recycled paper. According to a proposal by [7] environmentally significant behaviour could be classified from an" intent-oriented" or an "impact-oriented" perspectives. Intent-oriented classification means that environmentally significant behaviour is defined by motivation of the actor. This is based on whether a particular behaviour is being taken by the actor with the intention to benefit the environment.

Although [9] posited that some behaviour is performed because they are perceived to be environmentally beneficial, therefore they do not necessarily result in a reduction of the actual impact on the environment. For impact-oriented induced behaviour, they do not focus on the motivation of the actor to perform certain behaviour but rather defines behaviour by its actual impact on the environment. Examples for measure of impact-defined behaviour are energy use, water use, or waste production [10]. This means that certain behaviours may oftentimes performed due to individual characteristics. As observed by [11] through the theory of Planned Behaviour, attitude makes the most impact on person's behaviour only when the existing conditions are favourable. What this suggests is that despite individuals being influenced by attitudes, still certain behaviour may not sometimes be performed because of personal characteristics. However, attitudes stimulate behaviour, influenced by personal, social or informational factors which help evaluate behaviour positively or negatively. For instance, if someone does have knowledge of or perceive conservation behaviour such as recycling as too difficult, it is less likely that such behaviour will be carried out.

In the case of environmental knowledge, reasonable documented evidence have however shown that variance in behaviour has been explained by environmental knowledge from 6% to 8% [12, 13, 3] indicating that environmental knowledge could have considerable explanations on the relational differences between the conservation behaviour and environmental attitudes. However, [3] noted that contentious statistics have not yet reconciled the impact direct or indirect environmental knowledge has on conservation behaviour while [12] posited that most of the studies on environmental knowledge have only investigated or at most two forms of environmental knowledge and as such it does not give a detailed analysis of the relative effects of various knowledge on behaviour. Therefore in a bid to reduce the over concentration on general knowledge of behaviour, Kaiser et al as cited by [3] articulated the following three forms of knowledge which operate together in promoting environmental behaviour. They include: (a) System Knowledge (b) Behaviour-related Knowledge (c) Effectiveness Knowledge.

The understanding of the above forms explains that system knowledge is generally linked to knowledge about the ecosystems and its processes and can be ascribed to be vital aspect of environmental knowledge. The behaviour-related knowledge which could be described as action-related knowledge is described as the ability to know what to about any existing environmental problem. This form of environmental knowledge is associated to a superior level of knowledge. Finally, the knowledge about the benefits of environmentally dependable actions is considered as where consumers have to choose from various probable actions to achieve the maximum environmental benefits.

This study believes that this form of knowledge (Effectiveness) is not against the doctrine of environmental motivations that brings about environmental benefits. This form of knowledge has been considered by [12] as most effective environmental knowledge. In his illustration to support this, [3] noted that to start recycling at home, an individual should have some understanding about the material streams. Moreover, a person must be familiar with nature in order to know effective way to overcome any environmental challenge. Having knowledge and understanding of what he or she wants to do or the understanding of nature could motivate him or her into being environmental friendly.

3. Review of Theories

3.1 Theory of Planned Behaviour (TPB)

The theory of reasoned action(TRA) and its subsequent developed version theory of planned behaviour(TPB) as articulated by [14] in 1991 is based on the perceived behavioural control component to account for behaviours that occurs without a person's volitional control and norms [15,16] . The proponents of this theory had argued that subjective norms refer to the strength of normative beliefs so also the motivation to comply with these beliefs, social and moral values. As observed earlier, the theory of planned behaviour is an extension of theory of reasoned action (TRA) by its addition of influences on behaviour beyond people's control. TPB theorists believed that for these influences to be measured by means of the perception of one's control, two assumptions have to be made (1) the predicted behaviour must be at least partially beyond volitional control (2) perception of control must reflect actual control upon behaviour with some accuracy [17].

The first assumption here has been claimed to be domiciled in the environmental and ecological domain while the second has been noted to be a possible imperfection of the planned behaviour approach [18]. Against this backdrop, the theorists of TPB have argued that actions contribute towards environmental preservation and/or conservation. Hence environmental attitude is singled out as one of the promising concept [19, 20]. Therefore this study is of optimistic view that the usefulness of attitude could be a better predicator of environmental and ecological behaviours. Attitude towards the environment and attitude towards ecological behaviour rooted in the framework of Fishbein & Ajzen tradition of attitudinal research signifies that is either the object of one's attitude is the natural environment itself or some aspects of it such as air quality or the attitude object is environmental or ecological behaviour such as recycling. Hence attitude towards the environment could be described as an environmental concern [21]

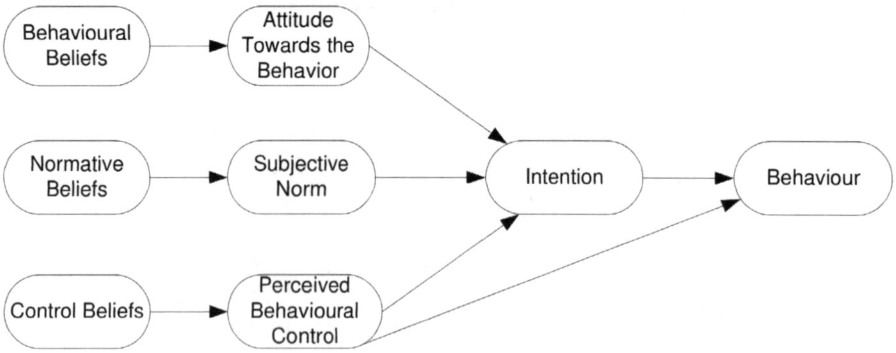

Figure 1. Theory of Planned Behaviour [14]

3.2 Theory of Value-Belief Norm (VBN): The premise of value-belief-norm theory (VBN) is that pro-social attitudes and personal moral norms are significant predictors of pro-environmental behaviour [6]. What this suggests is that, people who undertake environmental action have at least some altruistic or moral reason for doing so, or they are moved by self-serving interests [22]. The theory of VBN according to Stern in his model starts from values, which form the basis for belief, which reciprocally underlies norms which dictate behaviour. This shows that depending on values (altruistic, biospheric, egoistic) one may more or less likely to accept that his or her behaviour impacts on the environment.

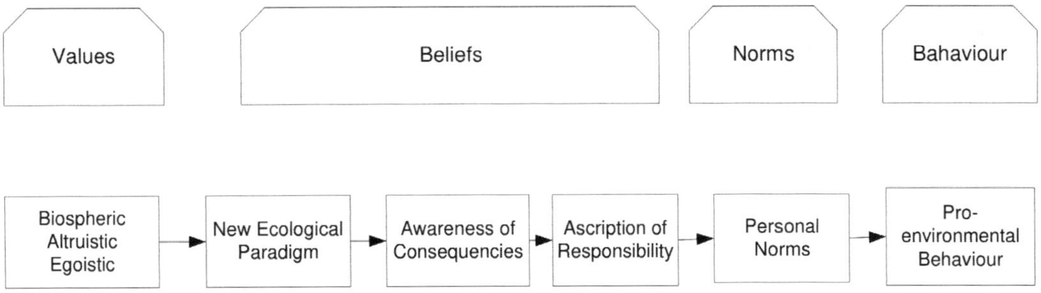

Figure 2. Value-Belief Norm Theory [7]

Each of the above theories has relatively high strength on environmental behaviour and as such they share similar explanatory features and characteristics that complement each other, hence this study found them very useful and attractive.

4. Environmental Motivations of Clients for Green Building Demand and Investment: A Personal and Altruistic Motivation Values

4.1 Quest for Enhancement and Protection of Eco-system and Bio-diversity: Theories of pro-environmental behaviour especially those related to protection and enhancement of eco-system and bio-diversity of environment and resources conservation in building such as the Value-belief Norm (VBN) theory assumes that the effect of values on behaviour is mediated through basic belief and pro-environmental personal norms. VBN acknowledges that both general altruistic norms cantered self-transcendent values that convey feelings of collective ownership of environmental welfare and personal moral norm based on self-enhancement or individual self-serving interests play significant role in green consumption and demand [23]. For example, as the growing sustainability and environmental awareness and consciousness continues to grow prospective home buyers, workers, tenants and investors will develop pro-environmental beliefs that could be attributed to altruistic or personal moral norms and values [23]. This means that some consumers may be demanding and investing in green properties not only because of its environmental, social and economic benefits such as energy saving and profit making but because of their altruistic belief that climate change and its effects on man and environment are real and could be detrimental, so they have to go green to mitigate the impending effects.

The theoretical concept of Value-belief Norm (VBN) theory as postulated by the theorists explain that if altruistic norms and personal moral norms rooted on the desire for value permeate deeply in consumers and clients, green consumerism, investment and sustainability will be a cardinal issue not only for those looking for new housing or office space but also for those renovating and retrofitting their building [23]. This will equally increase the demand for green properties. This is consistent with view that a little increase in environmental consciousness as articulated by VBN theorists could make home owners, office workers, tenants and prospective home buyers develop pro-environmental beliefs that are based on altruistic or personal norms and values. Proponents of this theory are of the view that one's subjective norms and normative beliefs regarding environment influence his or her intention to behave ecologically. What this suggests is that one's environmental value could make him or her go for green building. However, the result of this move could have weak or fair relationship as they could be presumably mediated by some economic, social and environmental variables [24, 25, 26, 27].

4.2 Quest for Conservation of Natural Resources: Motivations and intentions as recognized by the theory of environmental behaviour and concern such as the Theory of Planned Behaviour (TPB) are important determinants of consumer behaviour towards green building. For example, consumer's positive environmental concern, behaviour or perspective of green building may be influenced by a predicted destruction of natural resources or global warming as a condition for green consumption [28, 29]. While TPB recognizes intentions and motivations as environmental determining factor for green demand, this study however believes that environmental behavioural concerns could also be motivated by altruistic beliefs, personal norms and beliefs which lead to the development of attitudinal and habitual conservation of natural resources. Hence, the behavioural, attitudinal and intentional attributes of green consumers tend to evaluate the likely environmental consequences and benefits associated with the demand and purchase of green building before its consumption.

Therefore TPB theorists believe that such appraisal or evaluation could make the consumers pro or anti conservation of natural resources. For example, consumer's perceived level of self-involvement towards the protection of the environment may prevent them from engaging in environmentally friendly activities to conserve the natural resources sustainability components such as piedmont's earth, water, air and biologic resources as well as matters such as creek fortification, hillside ranking and management of the city urban forest etc. Since the methods of construction in green buildings are carried out to reduce and mitigate the impact on the environment, then many green building design principles help to condense impacts on natural resources and ecosystems. Some of the key examples identified by [30] include; sustainable sitting approaches which consider substitute to Greenfield construction, including using existing facilities(for example, urban redevelopment) and Brownfield sites, and avoid building on prime agricultural land, floodplains, and habitats for threatened species or near wetlands, parklands, and cultural or scenic areas. The philosophical quest also include designing to shrink potentially detrimental conditions, such as slope that can erode; avoiding adverse impacts on neighbouring properties; and carefully considering the building placement amid existing trees on site.

4.3 Quest for Reduction of Solid Waste: The broader literature of value belief norm theory on pro-environmental behaviour highlights the diversity of environmental and ecological factors that influence different environmentally significant behaviours which environmental values or concern could play a role. Consumption behaviour and adoption of new green products has been for example linked to identity [31, 32, and 33]. Self-identity has been found to be a significant predictor of behaviour over VBN variables. There exist two levels at which identity could operate in the context of pro-environmental behaviour [33]. They include: behaviour-specific and behaviour-generic. Behaviour specific include for example identity as a typical recycler [34] while behaviour generic encompass a sub-set of environmental actions such as green consumption and demand [35, 33]. Green architectural demand and consumption seeks to cut solid waste and materials used during construction. [36] observed that one simple way that green architecture creates less waste is by using renewable materials, such as plant matter and sustainable lumber, or reusing traditional materials, such as recycled stone or metal. Ecologically dependable green building can also help the occupants of the building waste fewer resources when going about their daily lives. Several green building design standards cut waste, which in turn minimizes the strain on landfills. In addition, using recycled materials in building construction encourages development of new industries that manufacture recycled products, further waste disposal needs and the use of virgin materials.

According to [30] the main sustainable design principles that decrease waste include the following (1) Storage and collection of recyclables: The building design it is hoped should offer space for collecting and storing materials such as paper, glass, plastic, and metals that will be recycled. (2) Construction waste management: At construction stage, the contractor can recycle or divert demolition and land-clearing waste from landfill disposal. (3) Recycled content: Designers can decide on environmentally favourable materials that comprise recycled materials. (4) Waste prevention: Designers can do away with needless finishes and make choices that use standard-sized or modular materials. In addition, designers consider product resilience in the design procedure. When products need to be changed less often, less demolition waste is created and fewer virgin resources are needed for substitution.

4.4 Quest for Enhanced Energy Efficiency and Improvement in Water and Air Quality: [29] had traced the research tradition of environmental attitude to two studies [37, 38] in consonance with TPB where factual knowledge about the environmental measures, the cognitive aspect, and verbal commitment measures the behaviour intention component of environmental attitude. This is why [39, 40] admitted that environmental and ecological behaviour appears to be susceptible to a wide range of influences beyond one's control. Using outside temperature and home characteristics as an example, the proponents of TPB theory noted that could affect energy consumption while cost of water affects water conservation and the number of people in a given household and house ownership [41, 42, 43, 44, 45]. Reduction in water utilization and protecting water quality are key objectives in green homes. Consumption of green building efficiently eases air pollution through condensed energy use, use of appropriate refrigerants, and use of materials with low off-gassing and other steps. The reduction in use of fossil fuel at the building site result in lower air pollution contributions at the site, while cut in electricity use results in lesser air pollution associated with power plants [46]. According to [46] energy efficiency is the keystone of every green home. In addition to saving homeowners money through lesser utility bills, an energy-efficient home may save some costs during construction.

4.5 Quest for Reduction of Societal Costs of Landfill Creation and Maintenance: Linked to the opinion of theorists of reasoned action and planned behaviour which proposes that attitude influences behaviour, mediated by intention, factual knowledge is seen as a precondition of any attitude, thus the relationship between factual knowledge and behaviour is mediated by intention [29]. However subjective norms and values are also mediated by intention and therefore could predict behaviour indirectly. Given these inter-related assumptions, motivation for green building by clients include a number of environmental safety and control rudiments in conformity with appropriate local, and state rules. The societal cost of landfill formation and maintenance could be guarded through the design and construction of green building. Cautious construction methods experts say can lessen the quantity of construction waste that get to landfills by 95% or more. The environmental issues of prime concern to the clients and to the community relative to land filling operation according to US EPA [47] include; water pollution, air pollution; land management and conservation and hazards and loss of amenity.

(1) *Water Pollution:* Ground and surface water can be contaminated by organic leachate from landfill sites. (2) *Air Pollution:* unrestrained landfill gas emissions are not a sustainable landfill practice. Landfills mostly produce methane and carbon dioxide which if not restricted can add to the greenhouse effect. (3) *Land Management and Conservation:* All land is precious and the impact of its use as landfill needs to be sustainable. Hence the motivational and appropriate care of a landfill through green building development as a valuable asset should end in resourceful remediation, allowing land to be used for other purpose following closure. (4) *Hazard and Loss of Amenity:* The possible risk and amenity impacts from landfills include fire, birds, dust, odour, pests, vermin and litter. Each of these probable impacts may happen on-site or off-site.

4.6 Quest for Minimization of Site Impact: One's subjective norms and normative belief regarding the environment as proposed by theory of planned behaviour influence his or her intention to adopt pro-environmental behaviour and attitude which could spur the desire to minimize site impact. These influence scholars believe ranges from weak to fairly large relationship to environmental beliefs and values [26, 27]. The relationship could reduce if ecological behaviour instead of ecological behaviour intention is considered [27, 21]. Theorists acknowledge that one's environmental values are related to ecological behaviour intention and could be mediated by other environmental behavioural intentions [24, 25, 29].

Therefore this could suggest that the intention to demand for site of a building influence a broad array of environmental issues as well as other factors such as security, accessibility, and energy consumption, energy used for transportation needs of tenants for commuting, the impact on local ecosystems, and the use/reuse of existing structures and infrastructures [48]. Green building construction and operations can have wide direct and indirect impacts on the environment, society, and economy, which seek to put in equilibrium the needs of above areas by using an incorporated approach to complement needs and generate design solution. All materials carry embodied environmental effects, in that there are environmental and ecological cost as an effect of their creation and use through their life circle. Green building demand promotes the use of materials with lower embodied environmental and ecological load. Green building also naturally uses fewer materials, through efficient design and removal of needless finish materials (for example, many green buildings utilize uncovered structural materials, rather than casing these materials with a wall finish). Green building processes in boosting impact on the environment also support recycling in their operation. Experts have observed that buildings use resources (energy, water, raw materials etc), generate waste (occupant, construction and demolition), and release probably injurious atmospheric emissions.

5. Corporate Conscience Environmental Responsibility Motivations: A Subjective Norm

Social pressure or informational factors could help assess behaviour in a positive or negative perspective or either to give in or defy social pressure. As observed by [11], the same factors could also have an effect on individual's perceptions to perform behaviour or resist it. Consequently, social and personal factors have an undeviating influence on the behavioural motivations and intentions that aid to predict behaviour. Subjective norms in the green building demand and investment context reflects individual's perception of social pressure regarding his or her ability to perform behaviour. For example, green lifestyle which has turned out to be fashionable and socially desirable and attractive these days has formed belief about sustainable demand which have influenced not only behavioural intentions and motivations but also behaviour [49, 50, 51, 52, 3].

Researches in the behavioural sciences according to [53, 54, 55] suggests that a green building sustains: relationship to nature; sense of community and belonging; behavioural choice and control; opportunity for habitual exercise; significant change and sensory variability and privacy when needed. Social sustainability is stood on the social facets such as feelings of well-being, aesthetics, health and comfort, security and user contentment, suitable living environment and social integration [56]. Behavioural choice and control as articulated in the VBN theory and TPB has been found to support the features and attributes of buildings linked to well being of individuals such as ensuring ventilation, temperature, and having the ability to modify and adapt environments to suit personal needs and preferences; multiple behaviour settings to support different activities; technology to support mobility; ability to move easily between solitude and social engagement and spaces to support both [54].

Again, by the principles of theory of planned behaviour (TPB) and Value belief norm theory (VBN), the society motivational social attitude towards the impact of environment has been interpreted into demand conditions that affect corporate behaviour. TPB and VBN acknowledgment of the intentional and motivational factors of green building demand such as social responsibility is a panacea towards avoiding harmful activities towards the environment. This is because it is principally based on ecological footprints and addresses the social, economic and ecological issues of green building in the community context [57]. Green demand completes all the social and community responsibilities of consumers while projecting the Client's image in the society [4, 58].

6. Environmental Skills and Proficiency Motivations: A Perceived Environmental Behaviour

The theory of planned behaviour has by extension added another determinant to behaviour - the Perceived Behavioural Control (PBC). According [14] this refers to person's perception of ease or difficulty to perform behaviour. PBC affects both behavioural intentions, motivations and behaviour and explains how person perceives his ability to perform such behaviour, which is dependent not only on his or her attitudes and societal limitations, but also on personal beliefs about contributing to environmental problem solving [3]. However, most behaviour has been found to be a continuum that extends from total control to a complete lack of control [59]. The individual has total control when there are no restrictions of any type to the adoption of a given behaviour. Contrarily, there is complete lack of control if adoption of the behaviour requires opportunities, resources, or skills which may be lacking.

However, beliefs about skills, resources and opportunities have been viewed as the underlying perceived behavioural control. The notion of PBC according to [59] and [17] is similar to [60] concept of self-efficacy belief and to Triandis's 1977 concept of "facilitating conditions" in his model of inter-personal behaviour [61]. It tends to reflect personal beliefs as to how trouble-free or hard adoption of the behaviour is likely to be. According to [61] PBC should be an important mediator of effect of habit upon the behaviour because those individuals who have performed a given behaviour in the past should have developed a feeling of control over performance of this behaviour. Therefore green building demand and investment as rooted in the TPB and VBN involves the motivational beliefs and potential capacity to determine how resources, skills and opportunities could be used in achieving the environmental expectations and acknowledgement of responsibilities by individuals. For instance, if clients and consumers have static expectations anchored on established motivating drivers such as skills, resources and opportunities, green consumerism and demand will be a lifestyle.

Table 1: Environmentally Theoretically Based Motivations and Expectations of Green Building Demand and Investment

S/N	Motivating Drivers	Expectations	Related Theory
1	Personal & Altruistic Environmental Motivations. (Values & Beliefs)	Ecologically induced green building demand is a claim to (1) protection of eco-system & bio-diversity (2) improved water & air quality (3) reduction of solid waste (4) conservation of natural resource (5) reduction of costs of landfill creation and maintenance (6) minimization of site impact (7) reduction to emission to air and energy efficiency enhancement	Theory of Planned Behaviour (TPB) Theory of Value Belief Norm (VBN)
2	Corporate Conscience Environmental Responsibility Motivations. Social Pressure Effect (Subjective Norms)	Green Building demand and investment by clients is seen as a social responsibility of caring for the environment while it stands as a response towards avoiding harmful activity towards environment rooted in environmental behaviour and principles.	Theory of Value Belief Norm (VBN) Theory of Planned Behaviour (TPB)
3	Environmental Skills and Proficiency Motivations. (Perceived Behavioural Control)	Green building requires investment capacity and green skills, resources and opportunities which are the underlying perceived behavioural controls.	Theory of Planned Behaviour (TPB)

7. General Framework and Theoretical contributions

A study of theory of reasoned action (TRA) and its subsequent developed version, the theory of planned behaviour (TPB) revealed that behaviour intention to perform is an immediate precursor of obvious behaviour while intention in turn is viewed as a function of one's attitude towards performing a particular act and perception of the expectations of relevant others (ie one's subjective norms). Again because attitude involves not just the evaluation of certain outcome but also the estimation of the likelihood of this outcome (expectations), hence factual knowledge is imperative and a pre-condition for any attitude [62, 29]. Apparently, as subjective norms demote the strength of normative beliefs and the motivation to fulfil these beliefs, social and moral values. What this means is that for one's subjective norms to be regarded as one, his or her normative stance, social expectations as well as moral values are considered approximately in relation to subjective norms. In the Value Belief Norm Theory (VBN) values are expected to exert a motivating influence on the perception, norms and actions of individuals, hence self-transcendent values could therefore mediate through pro-environmental norms. The moral character of pro-environmental norms and the environmentally beneficial behaviour has effect on the individual, society and organization. This gives credence to the assumption that the concepts related to organizational culture and the concept of theory of value belief norm are correlated [63].

The theory of Planned Behaviour (TPB) and the Theory Value Belief Norm (VBN) have received wide support in literature recently. For instance, the subjective norms was found to motivate a person to undertake socially desirable actions by [64] while [65] observed that consistent with the TRA as modified by TPB, green purchase and consumerism are significant predictors of green demand behaviour. Again [66] through the adoption of the above theory predicted the consumer purchase of organic foods usage among Malaysian consumers. Therefore since these theories assume that intentions and motivations are functions of attitudes and subjective norms, and they are the immediate antecedents of behaviour. It however extends TPB and VBN to recognize that environmental factors and perceived behavioural control such as skills, abilities, resources and opportunities can moderate and mediate the intention/motivation-behaviour relationship. This furthers underscores the focus of this study on the use of TPB and VBN to validate environmental motivations of green demand and investment. Hence the graphical presentation of framework for environmental motivations and expectations of green building demand and investment is depicted below for further explanations.

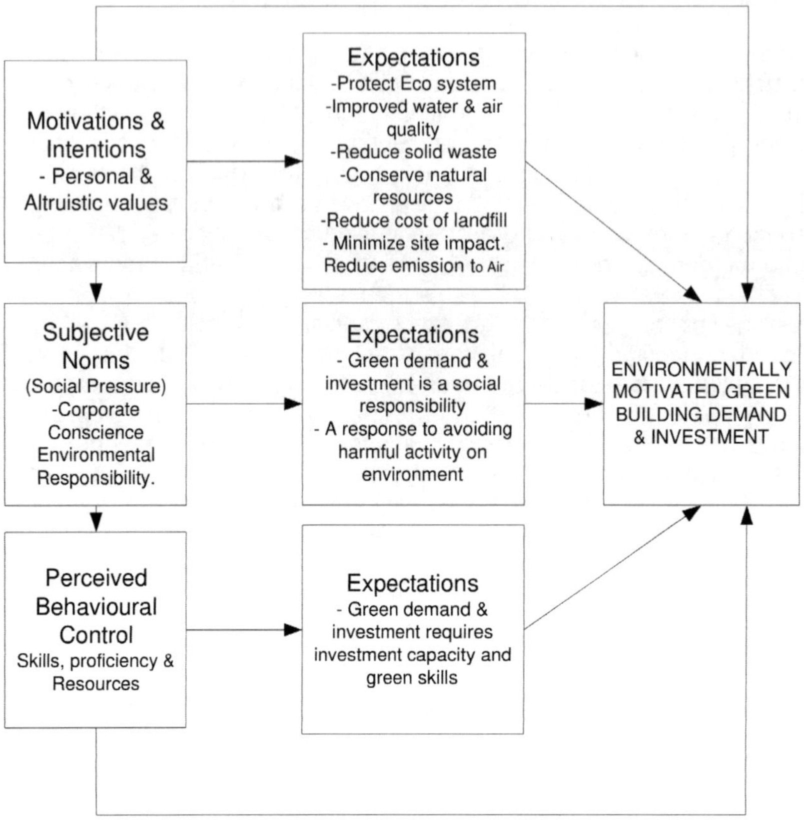

Figure 3. Framework for Environmental Motivations and Expectations of Green Building Demand and Investment

8. Barriers to Environmental Green Building Demand and Investment

The environmental and ecological desire to be green compliance in the building industry is often characterized and bedevilled with some challenges and barriers. Sometimes these problems are regulatory and non-regulatory, real and perceived. One of the major barriers to the growth of environmental sustainable building is lack of interest and awareness of green dividends and motivations on the part of the client. Awareness refers to clients' knowledge of the potential benefits of green buildings to him and his environment. Lack of knowledge, education and training in sustainable design, construction and demand underscores the failure of properties to account for the recovery of long-term savings, and the higher costs of sustainable building options [67]. Other barriers that could hinder the motivational ability of clients to go for green is lack of understanding of the sustainable property development [68]. [28] & [69] argued the common problems confronting the performance of green building industry to include: low level of education, people's resistance to change, limited post occupancy evaluation, greening existing buildings, lack of transit oriented development, separate capital and operating budgets, split incentive for owner-tenant and above all lack of integrated design. Additionally, laziness or inconveniences have been identified as the most common reason some clients are not acting in more sustainable way [70]. Lack of availability and poor current technology to satisfactorily meet the desired green and environmental building compliance standard is a major challenge. This limits the availability of higher performance green building components because certain green products are designed to meet the environmental conditions of a place. For example, the absence of high efficiency hot water tanks in some buildings could derail green construction. This means that builders and developers cannot use the most energy efficient or environmentally low impact products for their specific region or climate. Local capacity to design and construct certain environmental compliance green products such as photovoltaic system is lacking in some countries [71]. Hence developers' knowledge of the relevant green building technologies, products and practices is very critical to meeting the environmental demand of green building.

Others barriers are ineffective environmental laws and rules, lack of functional building codes to accelerate green construction and development. Most environmental laws in some countries are not ecologically friendly and this has increased both individual and organizational disinterest in environmental sustainability while project cost escalation has remained constant. Poor and lack of incentives to drive green building construction has remained a source of concern to developers. In some countries there exist little or no financial incentives such as tax credit/abatement, fee reduction/waiver, grants and loans as a means of encouraging developers. In some cases, most government offers monetary incentives with high interest rates and stringent conditions attached. According to [71] and [23] lack of incentive from government towards developers, increase of the maintenance cost, old age of existing building, weak affordability of extensive roof to withstand wind load, weak structural loading for applying extensive green roof system to resist environmental pressure, poor utilities arrangement, and lack of awareness on extensive green roof system in public and private sectors are some of the existing challenges. Hence effective allocation of resources as well as monitoring and controlling and understanding of green building objectives and possessing the ability to solving the problems will aid in the growth of sustainable green building.

9. Conclusion

Through the adoption of theoretical concept of Theory of Planned Behaviour (TPB) and Theory of Value Belief Norm (VBN), this study has examined the environmental behavioural motivations, intentions and beliefs, values and norms that could influence demand for green building investment with particular focus on the environmental motivating drivers and expectations. The study concludes that conservational behaviour of clients to demand green building is influenced by behavioural motivations and intentions. Behavioural motivations and intentions exhibited by clients and consumers of green building often reflect in their behavioural differences.

The study also found out that in consistency with perceived behavioural control (PBC), social pressure arising from institutions, family, friends and the society at large impacts on the behavioural motivations and intentions of green consumers. The study is therefore of the view that consumers who experience high social pressure are likely to be environmentally friendly than those with frail pressure. This underscores that perceived behavioural control as articulated in TPB influence not only behavioural intentions and motivations of green building consumers but affects their actual behaviour. Again, environmentally friendly attitudes were found out to play a significant role in behavioural intention and motivations which in turn is also influenced by perceived behavioural control. Therefore the society, the family and institutions can together effectively harnessed the expectations of the perceived behavioural motivations and intentions clients as grounds towards playing a role in the evolution of green consumerism and the behaviour around it.

References

[1] IMF Working Paper 2011 Who's Going Green and Why? Trends and Determinants of Green Investment. **38 - 40**
[2] Juraite K 2002 *Environmental Consciousness and Mass Communication: Construction of Public Opinion on the Environment in the Mass Media.* (Vytautas Magnus University Kaunas)
[3] Andrius N 2013 Using the Theory of Planned Behavior to Investigate the Determinants of Environmental Behavior among Youth. *Environmental Rsearch, Engineering and Management,* **1-63** 74 - 81
[4] Nurul A D and Zainul N A 2013 Motivation and Expectation of Developers on Green Construction: A Conceptual View. *World Academy of Science, Engineering and Technology* **76** 4 - 27
[5] Linda C J, Nisreen B, Mayuresh K and Nicole C 2011 Walking the Walk: How the Theory of Reasoned Action Explains Adult and Student Intentions to Go Green. *Applied Business Research* **27** 3
[6] Stern C P, Thomas D, Abel T, Guagnano A G, and Kalof L 1999 A value-belief-norm theory of support for social movements: The case of environmentalism. *Human Ecology Review* **6** 81 - 97
[7] Stern P C 2000 Towards a Coherent Theory of Environmentally Significant Behaviour. *Social Issues* **56** 407 - 424
[8] Stern P C, Dietz T, Ruttan W V, Socolow H R, and Sweeney L J 1997 *Environmentally Significant Consumption.* (Washington, DC: National Academy Press)
[9] Wouter P, Linda S, and Charles, V 2004 Values, Environmental Concern, and Environmental Behavior : A Study Into Household Energy Use. *Environment and Behavior* **36** 70 - 93
[10] Gatersleben B, Steg L, and Vlek C. 2002 The Measurement and Determinants of Environmentally Significant Consumer Behaviour. *Environment and Behavior* **34** 335 - 362.
[11] Ajzen I, and Fishbein M 2005 *The Influence of Attitudes on Behavior.* In D Albarracin, B. T. Johnson, and M. P. Zanna *(Eds). The handbook of attitudes.*
[12] Kaizer F G, Frick J and Wilson M 2004 Environmental Knowledge and Conservation Behavior: Exploring Prevalence and Structure in a Representative Sample. *Personality and Individual Differences* **37** 1597 - 1613
[13] Hwang Y H, Kim I S, and Jeng J M 2000 Examining the Causal Relationships Among Selected Antecedents of Responsible Environmental Behavior. *Environmental Educatio* **31** 19 - 25.
[14] Ajzen I 1991 The Theory of Planned Behavior: Organisational Behavior and Human Decision Processes **50** 179 - 211
[15] Ajzen I and Fishbein M 1980 *Understanding Attitudes and Predicting Social Behavior:* (Englewood Cliffs NJ: Preentice Hall)
[16] Ajzen I, and Madden J T 1986 Prediction of Goal-Directed Behaviour: Attitude, Intentions, and Perceived Behavioural Control. *Journal of Experimental Social Psychology* **22** 453-474
[17] Kalafatis P S, Pollard M and Tsogas H M 1999 Green marketing and Ajzen's theory of planned behavior: A cross-market examination. *Consumer Marketing* **16** 441- 460.
[18] Newhouse N 1990 Implications of Attitude and Behavior Research for Environmental Conservation. *Journal of Environmental Education* **22** 26 - 32
[19] Axelrod J L, and Lehman R D 1993 Responding to Environmental Concern: What Factors Guide Individual Action? *Journal of Environmental Psychology,* **13** 149 -159.
[20] Vining J and Ebreo A 1992 Predicting Recycling Behavior from Global and Specific Environmental Attitudes and Changes in Recycling Opportunities *Journal of Applied Social Psycholog* **22** 1580 -1607
[21] Katrin P 2012 Understanding Behaviour Chnge: How to apply theories of behaviour change to SEWeb and related public engagement activities. *Scotland's Environment* **4**
[22] Aliagha U G, Maizon H, Afeez S O, and Ali K N 2013 Review of Green Building Demand Factors for Malaysia. *Journal of Energy Technologies and Policy* **3** 11
[23] Dunlap E R and Liere D K V 1978 The New Environmental Paradigm: A Proposed Instrument and Preliminary Results *Journal of Environmental Education,* **9** 10 -19
[24] Axelrod J L 1994 Balancing Personal Needs with Environmental Preservation: Identifying the Values that Guide Decisions in Ecological Dilemmas. *Journal of Social Issues* **50** 85 -104
[25] Midden H J C and Ritsema M S B 1983 The Meaning of Normative Processes for Energy Conservation. *Journal of Economic Psychology* **4** 37- 55

[26] McGuinness J, Jones P A and Cole G S 1977 Attitudinal Correlates of Recycling Behavior. *Journal of Applied Psychology* **62** 376 - 384
[27] Aliagha U G and Yin N C 2013 Perceptions of Malaysia Office Workers on Adoption of Japanese Cool Biz Concept of Energy Conservation *Journal of Asian and African Studies* **48** 427- 446
[28] Kaiser F G, Wolfing S and Fuhrer U 1999 Environmental Attitude and Ecological Behaviour. *Journal of Environmental Psychology* **19** 1-19
[29] William B 2003 *The Environmental Benefits of Sustainable Design* (Building Green Inc)
[30] Cook J A, Kerr N G, and Moore K 2002 Attitudes and Intentions Towards Purchasing GM Food *Journal of Economic Psychology* **23** 557- 572
[31] Grewal R, Mehta R and Kardes R F 2000 The Role of the Social-identity Function of Attitudes in Consumer Innovativeness and Opinion Leadership. *Journal of Economic Psychology* **21** 233 - 252
[32] Lorraine W and Saffron O 2010 Green Identity, Green Living? The Role of Pro-environmental Self-Identity in Determining Consistency Across Diverse Pro-environmental Behaviours *Journal of Environmental Psychology* **30** 305 - 314
[33] Mannetti L, Pierro A and Livi S 2004 Recycling: Planned and Self-expressive Behaviour *Journal of Environmental Psycholog* **24** 227- 236
[34] Sparks P and Shepherd R 1992 Self-identity and the Theory of Planned Behaviour: Assessing the Role of identification with Green Consumerism *Social Psychology Quarterly* **55 4** 388 - 399.
[35] Jason K 2011 The Human Benefits of Green Building. Retrieved from http://sustainablecitiescollective.com/jasonking/20484/guest-post-h..
[36] Maloney P M and Ward P M 1973 Ecology: Let's hear from the people. An objective scale for the meaurement of ecological attitudes and knowledge *Journal of American Psychologist* **28** 583 - 589
[37] Maloney P M Ward, P M and Braucht N G 1975 Psychology in Action: A Revised Scale for the Measurement of Ecological Attitudes and Knowledge *Journal of American Psychologist* **30** 787- 790
[38] Smythe C P and Brook C R 1980 Environmental Concerns and Actions: A Social-Psychological Investigation. *Canadian Journal of Behavioural Science* **12** 175 -186
[39] Hines M J, Hungerford R H and Tomera N A 1986/87. Analysis and Synthesis of Research on Responsible Environmental Behavior: A Meta-Analysis *Journal of Environmental Education* **18** 1 - 8
[40] Olsen E M 1981 Consumers' Attitudes Towards Energy Conservation. *Journal of Social Issues* **37** 108 -131
[41] Verhallen M M T and Raaij F W V 1981 Household Behavior and the use of Natural Gas for Home Heating. *Journal of Consumer Research* **8** 253 - 257
[42] Moore S, Murphy M, and Watson R. 1994 A Longitudinal Study of Domestic Water Conservation Behavior. Population and Environment *Jouranal of Interdisciplinary Studies* **16** 175 -189
[43] Gamba J R and Oskamp S 1994 Factors Influencing Community Residents' Participation in Commingled Curbside Recycling Programs *Environment and Behavior* **26** 587- 612
[44] Lansana M F 1992 Distinguishing Potential Recyclers from Nonrecyclers: A Basis for Developing Recycling Strategies. *Journal of Environmental Education* **23** 16 - 23
[45] Green-Homes 2013 A Green Building Overview: Minimizing The Residential Sector's Toll on The Environment with 10 Practices Retrieved from http://www.hgtvremodels.com/interiors/a-green-building-overview/i.
[46] EPA 1996 *Environmental Guidelines: Solid Waste Landfills* (US Environmental Protection Agency)
[47] NIBS 2013 *Optimize Site Potential.* Retrieved December, 10th, 2013
[48] Riethmuller S H and Buttriss G J 2008 *Closing the gap between Pro-environmental Attitudes and Behaviour in Australia.* Paper presented at the Proceedings of the Australian and New Zealand Marketing Academy ANZMAC.
[49] Kaizer F G and Gutscher H 2003 The Proposition of a General Version of the Theory of Planned Behavior: Predicting Ecological Behavior *Applied Social Psychology* ***33*** 3 586 - 603.
[50] Kaizer F G and Scheuthle H 2003 Two Challenges to a Moral Extension of the Theory of Planned Behavior: Moral Norms and Just World Beliefs in Conservationism. Personality and Individual Differences *Personality and Individual Differences* ***35*** 1033 - 1048.
[51] Valle P, Rebelo E, Reis E and Menezes J 2005 Combining Behavioral Theories to Predict Recycling Involvement *Environment and Behavior* **37** 3 364 - 396
[52] Boyden S 2004 *The Biology of Civilization: Understanding Human Culture as a Force in Nature* (New South Wales, University of New South Wales Press)
[53] Orians G H and Heerwagen J H 1992 *Evolved Responses to Landscapes. In The Adapted Mind: Evolutionary Psychology and the Generation of Culture* (Oxford and New York: Oxford University Press)
[54] Heerwagen J H and Orians H G 1993 *Human, habitats and aesthetics. In S R Kellert and E O Wilson(Eds) The Biophilia Hypothesis* Washington DC Island Press Shearwater Books.
[55] Berit S 2010 *Impact of Sustainability on Property Values* (University of Regensburge Germany)
[56] Kibert C J 2007 *Sustainable Construction; Green Building Design and Delivery* (New York John Wiley and Sons)

[57] Chan E H W, Qian Q K and Lam P T I 2009 The Market for Green Building in Developed Asian Cities - The Perspectives of Building Designers *Journal of Energy Policy* **37** 8 3061- 3070.
[58] Gaston G, Pierre V and Linda L 1993 The Pattern of Influence of Percieved Behavioral Control Upon Excercising Behavior: An Application of Ajzen's Theory of Planned Behavior *Behavioral Medicine* **16** 1
[59] Bandura A 1977 Self-efficacy Mechanism in Human Agency *Psychol* **37** 122 - 147.
[60] Ajzen I 1985 *From Intentions to Actions: A Theory of Planned Behavior.* . (Heidelberg, Germany Springer)
[61] Stutzman M T and Green B S 1982 Factors Affecting Energy Consumption: Two Field Tests of the Fishbein-Ajzen Model *Journal of Social Psychology* **117** 183 - 201
[62] Anders B and John T 2007 *The Interaction of Values and Norms to Promote Sustainable Consumption and Production.* Paper presented at the Ist International Workshop on Life Cycle Approaches to Sustainable Consumption Tokyo Japan.
[63] Golan J G and Banning A S 2008 Exploring a link between the Third-Person Effect and the Theory of Reasoned Action: Beneficial Aids and Social Expectations. *American Behavior Science* **52** 2 208 - 224
[64] Chan R K Y 2001 Determinants of Chinese Consumers' Green Purchase Behavior. *Psychology and Marketing* **18** 4 389 - 413.
[65] Shaharudin R M, Pani J J, Mansor W S and Elias J S 2010 Factors Affecting Purchase Intention of the Organic Food in Malaysia's Kedah State. *Cross-Cultural Communication* **6** 2 105 - 116.
[66] Landman M 1999 *Breaking Through the Barriers to Sustainable Building: Insights from Building Professionals on Government Initiatives to Promote Environmentally Sound Practices* (TUFTS University US)
[67] Perrett A G 2011 *The Key Drivers and Barriers to the Sustainable Development of Commercial Property in New Zealand (* Lincoln University)
[68] Bilau G 2008 *Eight Challenges Facing the Green Building Industry Official*
[69] Bond S 2011 Barriers and Drivers to Green Buildings in Australia and New Zealand. *Property Investment and Finance* **29** 4/5 494 - 509.
[70] Sheltair-Group-Innes-Hood 2013 *Green Residential Building in North America: The Benefits of a North American Strategy; A Perspective from Canada* (The Sheltair Group Innes Hood)
[71] Zhang X, Shen L, Tam Y W V and Lee L Y W W 2012 Barriers to Implement extensive Green Roof Systems: A Hong Kong Study *Renewable and Sustainable Energy Reviews* **16** 314-319

Implementing gravity method on geological contacts in Bukit Bunuh, Lenggong, Perak (Malaysia)

I N E Hidayah[1], Rosli Saad[1], Mokhtar Saidin[2], M M Nordiana[1], I N Azwin[1] and Andy Anderson Bery[1]

[1]Geophysics Section, School of Physics, 11800 Universiti Sains Malaysia, Penang, Malaysia
[2]Centre for Global Archaeological Research Malaysia, 11800 Universiti Sains Malaysia, Penang, Malaysia

Email: noerelhidayah@yahoo.com

Abstract. Using gravity method, a study of responses of different geological settings towards gravity was conducted in Bukit Bunuh, Lenggong, Perak (Malaysia). Gravity method provides different responses towards different rock types depending on the rock density. The survey area cover an area of 12 km^2, with a total 404 survey stations with stations interval of approximately 50 m and 500 m. The Bouguer anomaly map identifies the shallow granite rock with a value of -6 to -10.5 mGal while shallow limestone is -11 to -15 mGal. The sediment/overburden was identified with value of < -15.5 mGal. Few fractures are also identified and the boundary between limestone and granite rocks identified at the north part of the study area.

1. Introduction

The study implemented gravity method which involves in measuring the Earth's gravitational field at specific locations on the Earth's surface to determine the subsurface density variations [1]. A geological contact is the surface along which one rock touches another and a boundary between two rock units which identified on the basis of a textural, compositional, temporal difference or structural between rocks [2]. Drilling, application of geophysical methods or geological mapping can infer a geological contact. Common geophysical methods used to delineate these features are magnetic, gravity, seismic refraction and electrical resistivity methods. The study is to perceive the geological contacts' response using gravity method. As different rocks gives off different gravity value, gravity method was chosen to delineate the existence of different geological settings at the same place. Generally, granitic rock have higher gravity value if were to be compare with limestone rock. Granitic rocks are composed of minerals rich in silicon and aluminium (felsic) which is denser than minerals contains in limestone rock, thus resulting in higher gravity value. Furthermore, gravity method is cost and time effective where it can cover over large areas in a short time. By doing the desk study and visiting the interested survey area, the different geological settings of granitic and limestone were found existed at the same place (N 5.09112° E 100.9724°) in Bukit Bunuh, Lenggong, Perak (Malaysia).

2. General Geology

Predominantly the Lenggong valley comprise of few lithologies which are tefra dust, granitic rock and allvium (figure 1). The entire of Lenggong valley is dominated with granitic rock which is from

IOP Conf. Series: Earth and Environmental Science **23** (2015) 012011 doi:10.1088/1755-1315/23/1/012011

Jurassic end-Carbonaceous low era and originated from Bintang Range at the west of Lenggong [3]. Almost all of alluvium units in which the quatenary sediments contain of alluvium and tefra dusts are located along the river area. The topography of the Lenggong valley is exceeding 600 ft MSL. Situated at the east of Bukit Bunuh is Sungai Perak which flows from north to south and located at 61ft MSL while Bukit Bunuh consists of the older granitic rock. 200 million years ago at Mesozoic era the granite intrusion was happened all over Malaysia [4]. In Lawin, Bukit Jawa, Kampung Telemong and Kota Tampan, pebbles which is dominated by quartz and quartzite are found at the elevation 200-350 ft. At the lower zone of <600 ft, there exists colluviums which are characterized by a lots of fractures and fault zones. The rocks physical features exhibits that they are fractured and angled fractured comprise of chert, flint and agate.

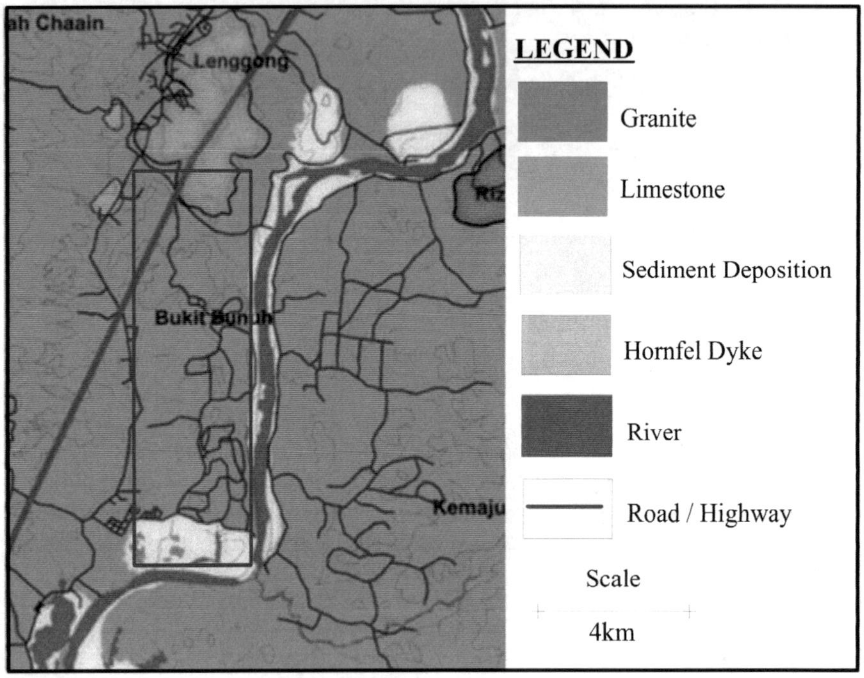

Figure 1. Geology map of Lenggong, Perak, Malaysia with study area marked in the purple box [5].

3. Theory of gravity method

Gravity method utilized to study the figure, composition and structure of the Earth. Density variations of bedrock and soil in the immediate vicinity of measuring points influence the force of gravity in a discernible way. Influenced primarily by the mass and figure of the Earth, local and regional topography and centrifugal force due to the Earth's rotation is total gravity [6].

Density is the parameter which control and determine the Earth's gravity value. The changes of the density value horizontally will change the gravity or the anomaly gravity horizontally. Thus, the gravitational attraction of a body of non-homogenous density will vary from point to point, in react to the distribution of density within the body. Due to the reason, measurements of the variation, with location of the gravitational attraction of the Earth can provide valuable information of subsurface geology [7]. Rock types and minerals have their own characteristics range of densities. The dispersal in the subsurface of the constituent rock types and minerals will reflect through changes in the local gravitational field. The variation of the gravitational field of the Earth from place to place may therefore, be interpreted in terms of subsurface geology – with certain assumption and limitations. Table 1 shows the density value for some of different materials [8]. In gravity anomaly interpretation, the measuring of the difference in density between different rock types should be done before determining its structure. The bulk density of the rock is controlled by the mineral content which forms rock mass, porosity and type of fluid in the cavity.

Table 1. Density values of different materials [8].

Materials	**Density , ρ (g/cm^3)**
Overburden	1.92
Granite	2.50 – 2.81
Limestone	1.93 – 2.90
Gravel	1.70 – 2.40
Shale	1.77 – 3.20
Sedimentary rocks	2.50

4. Data acquisition

Similar with any other geophysical survey, this study also started with desk study by studying the geological settings of the interested area. The lines and stations spacing were also decided depending on the area that need to be cover and the accessibility. After all of the decisions being made and all the preparations for the gravity survey were done, the gravity measurements were conducted using gravimeter Scintrex CG-5. The equipment measure differences in gravity from station to station [7] in miliGals (mGal). The survey conducted using loop principal to monitor and eliminate the potential drift effect. Base readings were taken at the beginning and end of the survey for that particular day at a fixed area where it is noise-free. The gridding survey stations were design to cover 12 km^2 areas. The northern traverse distance is 6 km with 50 m station spacing and eastern traverse distance is 2 km with 500 m station spacing. The total gravity stations for the study area are 404 (figure 2). Note that all of these stations are not taken on the same day but taken on different but continuous days. The global positioning system (GPS) instrument used for stations coordinates and stations elevations. The surveys were conducted in Bukit Bunuh, Lenggong valley (Perak) where meteorite impact was suspected million years ago [9]. After data acquisition, all the readings taken on that day will be transfer from the equipment to computer for data processing and interpretation. All the data collected were compiled and corrected using several standard procedures such as drift correction, elevations correction and free air correction to produce corrected Bouguer anomaly and contour map of Bouguer anomaly. Several filtering scheme were apply such as low pass filter and then the patterns of anomalies in contour map are observed and interpreted.

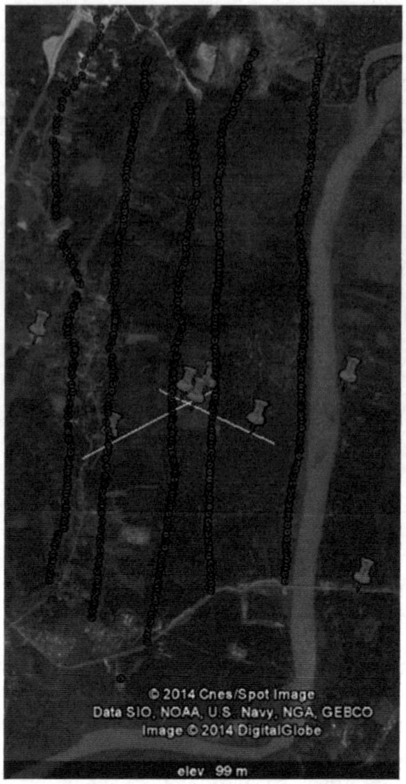

Figure 2. The survey lines and stations conducted and also borehole locations (red pins) in the survey area.

5. Result and discussions

Figure 3 shows Bouguer anomaly map of the study area with a value of -6 to -18 mGal. The high value (-6 to -10.5 mGal) is identified at northeast part and scattered at the center of the study area. This is due to the shallow level of granitic rock. The shallow level of granitic rock is validated from the borehole record which was done scattered in the survey area. The borehole records denoted granitic bedrock as impact granite with depth of 5-40 m. The rest of the study area was identified with a value of -11 to -15 mGal. The northern part is suspected due to the shallow limestone rock while the rest is due to sediment/overburden. Fractures also identified with the value of less than -15.5 mGal. The gravity low indicates rock fracturing as fractured rock is less dense than the unaltered rock. Previous studies conducted in the survey area using other geophysical methods such as seismic refraction, 2-D electrical resistivity method and magnetic method also show result with a lot of fractures zones due to the impact meteorite [10, 11; 12].

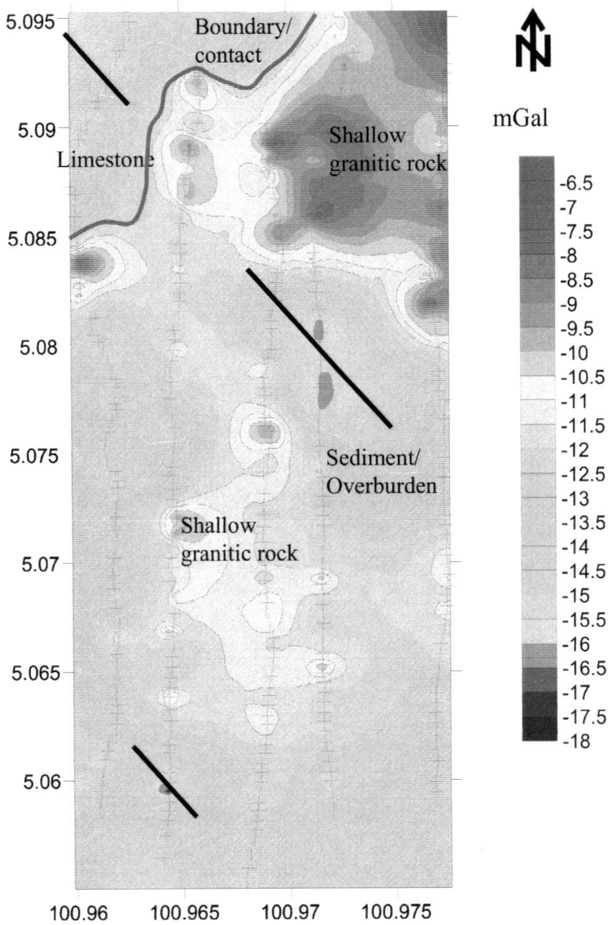

Figure 3. Bouguer anomaly map in Lenggong, Perak with fractures represented by the black lines.

6. Conclusion

Based on the Bouguer anomaly map produced, the difference of responses of different geological settings is significance which in this case granitic (-6 to -10.5 mGal) and limestone (-11 to -15 mGal) while fractures zone indicated with much lower value of less than -15.5 mGal. As the result shows distinct differences in the values between two different geological settings, it shows that gravity method is capable to map the study area with density respond and identifies shallow or deep bedrock, fractures and boundaries/contact.

7. Acknowledgments

The authors wish to thank the member of Geophysics group of USM for their cooperation in the geophysical field data collection and Centre for Global Archaeological Research (CGAR), Universiti Sains Malaysia for sponsoring the project.

References

[1] Reza T 2010 Application of gravity method in fault path detection *Aus. J. B. & Appl. Sc.* **4(12)** 6450-60

[2] Howe R C 1997 Geologic contacts *J. Geosc. Edu.* **45** 133-6

[3] Mokhtar S 1993 Kajian perbandingan tapak Paleolitik Kampung Temelong dengan Kota Tampan dan sumbangannya terhadap kebudayaan zaman Pleistosein akhir di Asia Tenggara *Msia. Museum J.* **32**

[4] Alexander J B 1962 A short outline of the geology of Malaya with special reference to Mesozoic orogeny: crust of the Pasific Basin *Geophy. Mono.* **6** 81-6

[5] Geological Map of Malaysia 1985 Geological survey department of Malaysia

[6] Elo S 1997 Interpretation of the gravity anomaly map of Finland *Fin. Geophy. Soc.* **33(1)** 51-80

[7] Seigel H O 1995 A guide to high precision land gravimeter surveys Scientrex Limited

[8] Linsley R K, Kohler M and Paulhus J 1982 *J. Geosc. Edu.* **45** 133-6

[9] Mokhtar S 2006 Bukit Bunuh, Lenggong, Perak : Sumbangannya kepada arkeologi dan geologi negara *J. Arkeo. Msia* **19** 1-14

[10] Mark J, Rosli S, Mokhtar S, Kiu Y C and Andy A B 2013 Second stage subsurface study of meteorite impact crater at Bukit Bunuh, Malaysia using 2D electrical resistivity survey *Elec. J. Geotech. Eng.* **18** 1199-204

[11] Rosli S, Kiu Y C, Mokhtar S and Shyeh S K M 2013 Identifying subsurface features using magnetic residual in Bukit Bunuh, Perak, Malaysia *Elec. J. Geotech. Eng* **18** 1367-74

[12] Kiu Y C, Rosli S, Mokhtar S, Nordiana M M and Andy A B 2012 Characterization of Bukit Bunuh ground subsurface by 2D resistivity for meteorite impact study *Elec. J. Geotech. Eng* **17** 3575-83

Fracture Analysis of basement rock: A case example of the Eastern Part of the Peninsular Malaysia

A Shamsuddin[1] and D Ghosh[1]
University Technology PETRONAS, 31750, Tronoh, Perak, Malaysia

E-mail: anassash@gmail.com, drdeva@petronas.com.my

Abstract. In general, reservoir rocks can be defined into carbonates, tight clastics and basement rocks. Basement rocks came to be highlighted as their characteristics are quite complicated and remained as a significant challenge in exploration and production area. Motivation of this research is to solve the problem in some area in the Malay Basin which consist fractured basement reservoirs. Thus, in order to increase understanding about their characteristic, a study was conducted in the Eastern part of the Peninsular Malaysia. The study includes the main rock types that resemble the offshore rocks and analysis on the factors that give some effect on fracture characteristic that influence fracture systems and fracture networks. This study will allow better fracture prediction which will be beneficial for future hydrocarbon prediction in this region.

1. Introduction

Reservoir rocks such as carbonates, tight clastics and basement rocks are the important elements in the petroleum system. The aid of fractures' elements can give some effects on increasing or decreasing the reservoir performance. The factors that influence fracture characteristics remain as a challenge in the exploration and production area.Two of the factors that influence fracture characteristic are rocks types and also stress (tectonic). Different rock types will produce different marks such as fracture as responses to the internal and also external factors.

Fractures can be defined as planar or subplanar discontinuity which forms due to the external or internal stress (Haakon F, 2010). Fractures are usually found in brittle structures as their rock body accommodate the stress or deformation in showing the marks when they lost cohesion through fractures characteristics (George HDet. al 2012).Fractures can be defined in three types which are shear fractures and extension fracture joints orfissures as shown in Figure 1.0: Types of fractures.

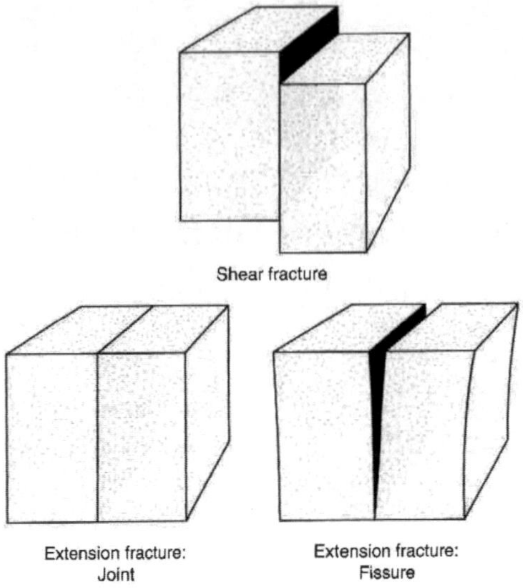

Figure 1.0: Types of fractures (Haakon F 2010)

Fracture analysis of the basement rocks is important as metamorphic and igneous fractured reservoir fields are already been discovered and developed in South East Asia including in the Bach Ho (Vietnam), Suban Gas Field (Indonesia) and Malay Basin (Malaysia). The definition of the basement rock is quite extensive as different researchers define the basement in many ways. According to Landes KK (1960), basement rocks are any metamorphic or igneous rocks which are overlain by sedimentary layer while North FK(1990) defines basement rocks to include any sedimentary origin if they have little to zero matrix porosity. In this paper, the researchers agree with Ngoc (n.d.) definition, as he defines the basement rocks are any metamorphic or igneous rocks which are overlain a study area of petroleum basin.

2. Geological setting of the study area

A study was conducted in the East domain of the Eastern part of the Peninsular Malaysia (Figure 2.0) as the rock types in this location resembles the basement rock in the offshore part and it can be a good analogue study. Based on the study, the main lithology of the research areas is granite and might be I-type granitoids as according to MetCalfe I (2013) I-type granitoids can be found in the Eastern Belts of the Malay Peninsula. Their characteristics are equigranular to weakly porphyritic with alkali feldspar-orthoclase to intermediate microline (Charles S H, 1977). This type of rock was chosen because Ngoc H Y (n.d) and Suhaileen S (2005) mention that there are various types of basement rocks including Mesozoic granitic (intrusive and volcanic rocks) in the Malay basin area.

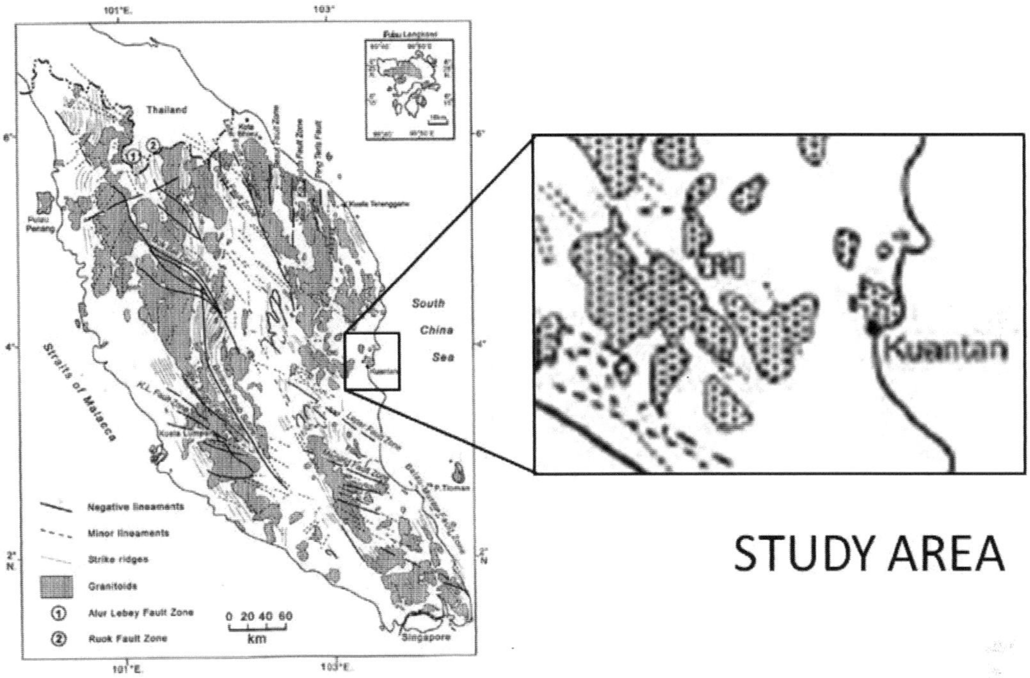

Figure 2.0: General geology of the Eastern part of the Peninsular Malaysia
(Charles S H and Tan D N K, 2009)

Figure 2.0 shows the granitoids bodies in the Peninsular Malaysia area.These granitoids bodieswith a few fault zones which are including Lebir fault zone, Besut fault zone, Kg Buloh fault zone, Ping Teris fault zone, Lepar fault zone, Mersing fault zone and Balau-Murau fault zone were mapped by Charles S H and Tan D N K (2009) become an interest as their fracture characteristic might resemble the offshore part. Three localities were chosen based on the good exposure of the fracture on the rock.

3. Methodology

Fracture analysis based on the scan-lines method was carried out on the three different localities. There were about eleven scan-lines with different length. The length of the scanlines was depending on the exposure of the rock. Six main data were collected in this location. Main data which based on discontinuity survey data sheet were collected. The data includes fracture types, fracture spacing strike, dip orientation and dip value. Distance and length of the coordinates (Figure 5.0) were collected by using Global Positioning System (GPS).

4. Rock types and Fracture Characteristics (Fracture Patterns)

Based on the study, three main features of the rocks' types determined are granite (the main type of rock), dolerite dykes and also enclaves. Generally, granites in this location compose main mineral like quartz (~1cm), feldspar (0.50mm-1.5mm) and plagioclase. Dolerite dykes can also bedetermined with black in colour features(Width: 0.06m-2m; Length:1m-20m). In certain part of the igneous rock, enclaves with a different mineralogy were determined. They might be fragment from the country rock.

Figure 3.0: Outcrop example of weathered granite with black dolerite dykes (left) and enclaves, xenoliths (right) in Teluk Cempedak Beach. Different types of mineral can be observed in this location

Granite in the study areas show fractures which are known as joints (Mode I). They are in ladder pattern which are in systematic joints and cross joints. Systematic joints are long and continuous and evenly spaced while the cross joints are evenly spaced joints which against the systematic joints (figure 4.0).

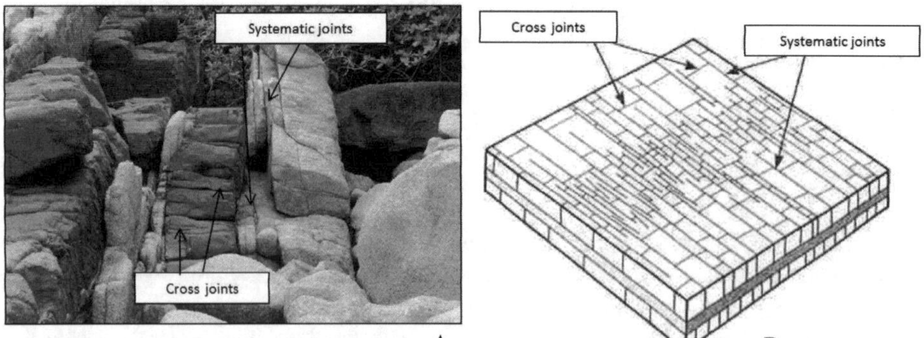

Figure 4.0: Two different joints which are systematic joints and cross joints in the outcrop in Teluk Cempedak (A) and comparison with George H D et. al 2012 fracture patterns (joints)

Based on the data analysis, there are two sets of fracture network for each of the locations except for the third location as this location is quite difficult to be assessed. Major fracture conjugates for the first location are NE-SW and NW-SE, the second location areNE-SW and E-W and third location is NE-SW.

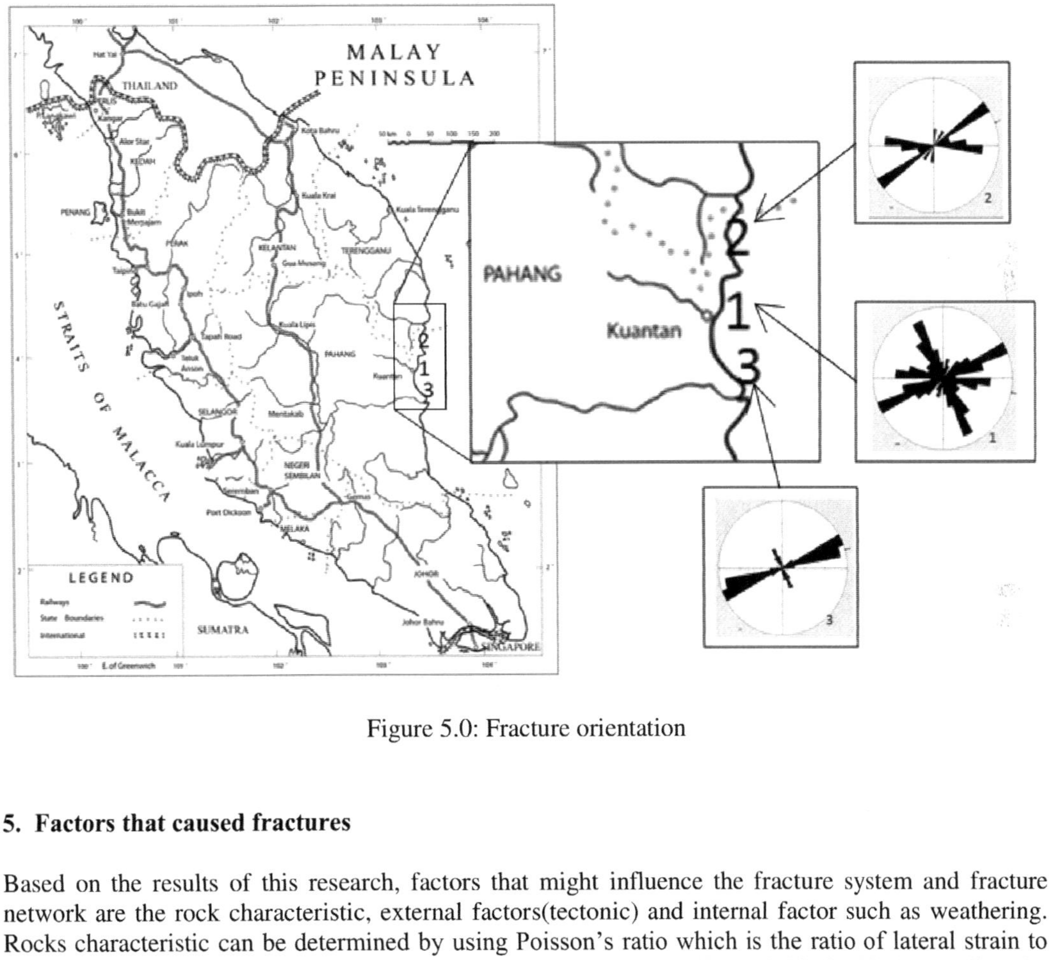

Figure 5.0: Fracture orientation

5. Factors that caused fractures

Based on the results of this research, factors that might influence the fracture system and fracture network are the rock characteristic, external factors(tectonic) and internal factor such as weathering. Rocks characteristic can be determined by using Poisson's ratio which is the ratio of lateral strain to longitudal strain. It is also known as extensional ratio (George HDet. al 2012). Understanding the characteristics of the rock especially in the value of Poisson's ratio (Figure 6.0) is important for characterising fractures.

$$= \frac{e\ \mathrm{lat}}{e\ \mathrm{long}}$$

Granite's poisson's ratio is about 0.04 (weathered) to 0.11 (fresh). This value shows that the granite tends to show brittle deformation and accommodate their pronounced loss of cohesion in showing some marks through fracture (joints) which is extension fin nature.

Rock type	v
Limestone, fine grained	0.25
Aplite	0.20
Limestone, porous	0.18
Limestone, oolitic	0.18
Limestone, chalcedonic	0.18
Limestone, medium grained	0.17
Limestone, stylolitic	0.11
Granite	0.11
Shale, quartzose	0.08
Graywacke, coarse grained	0.05
Diorite	0.05
Granite, altered	0.04
Graywacke, fine grained	0.04
Shale, calcareous	0.02
Schist, biotite	0.01

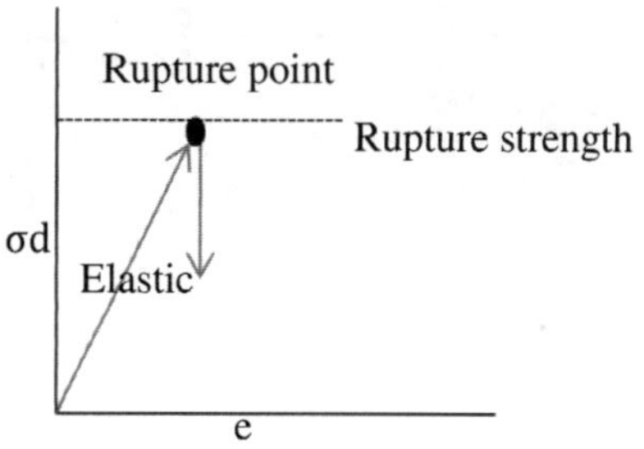

Figure 6.0: Poisson's ratio of granite compared to the other rocks (left) and their behaviour in the experimental deformation structures under extension and contraction (right) (Haakon F 2010)

6. Conclusion

Understanding rock characteristics and rock patterns like systematic joints and cross joints is crucial because they might be conduit or seal for the reservoir rocks. Other than that,it is important to identify fracture sets for each of the location as they might be influenced by local tectonics like in this area as there are a few set of fracture sets which are NE-SW and NW-SE for the first location, NE-SW and E-W for the second location and NE-SW for the third location is. Further study should be conducted to understand the fracture characteristic in spatial space and which tectonic factors that might influence them.As conclusions, fracture characteristic in the onshore (as an analogue) study might give some good predictions for the study on the offshore data.

7. Acknowledgement
I wish to thank PETRONAS and University Technology *PETRONAS* for research funding.

8. References
[1] Haakon F, 2010. *Structural Geology*
[2] George H D, Stephen J R and Charles F K, 2012*Structural Geology of Rocks and Regions*
[3] Landes K K, Amoruso J J, Charlesworth L J, Heany F, and Lesperance J P, 1960, Petroleum resources in basement rocks
[4] North F K, 1990*Petroleum Geology*, Unwin Hyman, Boston, 2nd edn, pp. 216–217.
[5] Ngoc HY,n.d. Seismic attributes for study Pre-Tertiary fractured basement in Malay Basin
[6] Metcalfe I, 2013. *Tectonic Evolution of the Malay Peninsula*
[7] Charles S H, 1977*Granite emplacement and tectonic subdivision of Peninsular Malaysia Geol. Soc. Malaysia*
[8] Suhaileen S, 2005 The prospectivity of Fractured Basement Play of the Malay Basin
[9] Charles S H and Tan D NK, 2009 *Geology of Peninsular Malaysia*

Combined analysis of 2-D electrical resistivity, seismic refraction and geotechnical investigations for Bukit Bunuh complex crater

I N Azwin[1], Rosli Saad[1], Mokhtar Saidin[2], M M Nordiana[1], Andy Anderson Bery[1] and I N E Hidayah[1]

[1]Geophysics Section, School of Physics, 11800 Universiti Sains Malaysia, Penang, Malaysia
[2]Centre for Global Archaeological Research Malaysia, 11800 Universiti Sains Malaysia, Penang, Malaysia

Email: nurazwinismail@yahoo.com

Abstract. Interest in studying impact crater on earth has increased tremendously due to its importance in geologic events, earth inhabitant history as well as economic value. The existences of few shock metamorphism and crater morphology evidences are discovered in Bukit Bunuh, Malaysia thus detailed studies are performed using geophysical and geotechnical methods to verify the type of the crater and characteristics accordingly. This paper presents the combined analysis of 2-D electrical resistivity, seismic refraction, geotechnical SPT N value, moisture content and RQD within the study area. Three stages of data acquisition are made starting with regional study followed by detailed study on West side and East side. Bulk resistivity and p-wave seismic velocity were digitized from 2-D resistivity and seismic sections at specific distance and depth for corresponding boreholes and samples taken. Generally, Bukit Bunuh shows the complex crater characteristics. Standard table of bulk resistivity and p-wave seismic velocity against SPT N value, moisture content and RQD are produce according to geological classifications of impact crater; inside crater, rim/slumped terrace and outside crater.

1. Introduction

The collision of meteorite with earth surface produce a bowl-like shape crater, disturbed its crust and altered its geological history [1]. It is easy to identify based on its unique shape however, earth processes cover the effect of the impacts, causing the crater to gradually fade away [2] and leave a very long-lasting, though not indelible, signature in soils and rocks [3]. The characteristics can be discovered using geophysical methods that efficiently imaged the earth subsurface. This study consists of field investigation and laboratory analysis. Field investigation employed two geophysical methods; 2-D electrical resistivity and seismic refraction methods and geotechnical boreholes drilling to identify the crater's type and its characteristics. It measures the resistivity, p-wave seismic velocity and number of blow in Standard Penetration, SPT N value. Soil and rock samples taken from boreholes are then analyzed in the laboratory to identify their moisture content and Rock Quality Designation (RQD) values. Based on characteristic shown from all the methods, standard parameter table is produced according to the geological classification; inside crater, rim/slumped terrace and outside crater.

2. General Geology and background

Study area is located at Bukit Bunuh, about 10 km from Lenggong town in the upper part of Perak state. Bukit Bunuh lies between two mountain ranges; Titiwangsa and Bintang Ranges and made up of Quaternary sediment and small lithology unit of Tertiary tefra ash and metasediments. The whole Lenggong Valley is underlain by granitic rock from Jurassic end to Carbonaceous low era which is originated from Bintang Range at the West of Lenggong [4]. Bukit Bunuh is suspected as meteorite impact crater due to more than 10000 surface suevite rocks associated with hypervelocity impact have been counted and accidently unearthed as a result of plantation activities. The suevites have been dated using the fission-track dating method by the Japan Geochronology Laboratory, Tokyo, Japan. The suevites are dated to be approximately 1.83 ± 0.61 million years ago that lies in the Quaternary period [5]. In order to register Bukit Bunuh as one of the world meteorite impact site and first to be discovered in tropical region, the boundary of the crater need to be verified accordingly.

3. Research methodology

Research methodology used in this paper comprises of field investigation and laboratory analysis. Field investigations are performed by utilizing 2-D electrical resistivity, seismic refraction and geotechnical borehole drilling methods. Samples taken from boreholes are tested for moisture content and RQD tests.

3.1. 2-D electrical resistivity

Resistivity method basically measures the resistivity distribution of the subsurface materials. Currents were injected through two current electrodes, C_1 and C_2 and potential difference at two potential electrodes P_1 and P_2 were measured. In this study, 41 units of stainless steel electrodes were connected to 4 ABEM Lund multi cables that lay in a straight line with 5m spacing resulting in 400m of total length in one spread. ABEM terrameter SAS 4000 and ABEM electrode selector ES 1064 were placed at the center.

Electrode configuration used in this survey was pole-dipole array where current electrode, C_2 was placed perpendicular and far from survey line and act as infinity. Roll along technique was also executed in order to cover the total length of survey lines. The inversion of field data was done using least-squares inversion scheme and processed using the model refinement cells with widths of half the unit spacing.

3.2. Seismic refraction

Seismic method measures the elastic wave that travelling through the subsurface. In this research, seismic refraction was performed using ABEM MK8 24 channel seismograph (display) with 24 units of 14Hz geophones (detector) and 40kg weight drop (source). Geophone spacing used is 5m resulting in 115m in one spread. Basically, geophones are connected to the seismic cable and planted the spike firmly into the ground. There are total of 5 inline shotpoints and 2 offset points. Survey continues by overlapping 1 geophone for the next spread until reached the total length of survey lines. In general, seismic wave is generated from seismic source, propagate downward through the subsurface and critically refracted when it encounter different density medium. Seismic refraction is based on the first arrival of signal that travels through a layer with higher velocity.

Critical refraction occurs when velocity of beneath layer is greater than velocity of the layer above it. Seismic refraction data is processed using tomography technique which known as velocity gradient or diving wave tomography. It uses first arrival travel time of seismic wave as input and uses the generalized simulated annealing where the algorithm performed repeated forward modeling, and new models are conditionally accepted or rejected based on probability criterion. This criterion allows the algorithm to escape from non-unique, local, travel-time minima to achieve a unique, globally optimized model of subsurface velocity structure.

3.3. *Standard Penetration Test (SPT) N value*

The SPT was carried out using the split barrel sampler and a self tripping hammer of 63.5 ±0.5 kg weights. The number of blows count for each 75 mm penetration of the sampling tube is known as N value. The test was carried out 1.5 m interval. The soil samples obtained from the split barrel were conserved as disturbed samples for laboratory analysis.

3.4. *Moisture content test*

Moisture content can be defined into two main basis which are weight basis or volume basis. Measurement of moisture content in the weight basis is known as gravimetric moisture content where the ratio of amount of water present in the void to the amount of solids is known as moisture content. However, it is often to express water content in the volume basis which is applied in this research that defined as the volume of water per bulk volume of soil.

3.5. *Rock Quality Designation (RQD)*

RQD is the ratio of the total length of good quality cores each exceeding 100 mm in length over the drilling. RQD is measured by taking the percentage of good rock sample from core barrel based on the quantity of intact pieces which is greater than 100 mm in length with the total length of the core barrel.

4. Survey lines and boreholes

The research covers approximately 8 km^2 assuming Bukit Bunuh as the center of the study area with Perak River separated the area between West and East part. The research divided into a three stages, starting with regional study consist of four survey lines with two lines of 8 km length oriented South-North (S-N) and West-East (W-E), 6.6 km lines cover the Northwest-Southeast (NW-SE) and 2 km line oriented Southwest (SW). Second stage focused on the West area of the Perak River with 5 parallel lines performed with total length approximately of 2 km (W1-W5). The research continues for the last stages that focus on the East boundary of study area. 12 survey lines were conducted with length varies from 200-1200 m (E1-E12). Total of 21 boreholes scattered in the study area. Figure 1 (a and b) shows the survey lines for three stages and location of boreholes respectively.

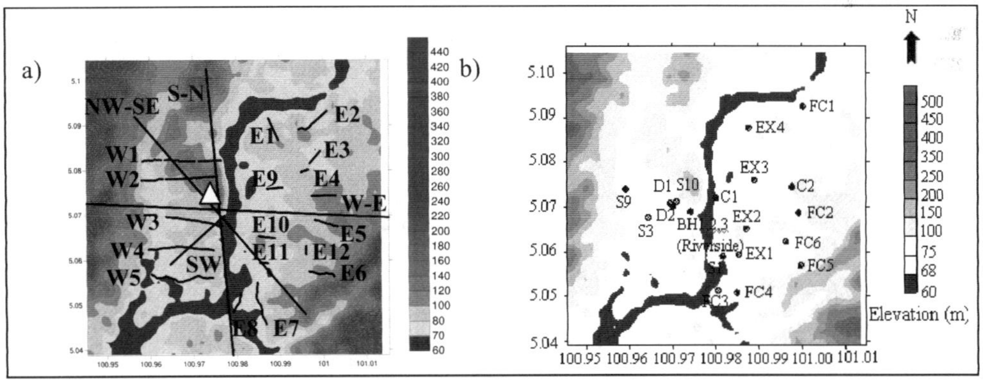

Figure 1. Survey lines and borehole location, a) survey lines for three stages and b) boreholes location of expected inside crater, on rim/slumped terrace and outside crater.

5. Results and discussions

Bedrock depths from 2-D electrical resistivity and seismic refraction results are digitized to produce bedrock topography contour map of Bukit Bunuh ancient impact crater. Bedrocks are digitized every 5-10 m based on resistivity and velocity of granite rocks at Lenggong area. Generally, bedrock depth varies from 5-50 m. Crater is found with diameter of 4.5-5 km surrounded with high elevated bedrock which defined as the crater rim consists of slumped terraces. Shallow level bedrock detected almost at the center which denoted as central uplifts or rebound zones. The Bukit Bunuh impact crater is classified as complex crater type. Figure 2 (a and b) shows the bedrock topography contour map of 2-D electrical resistivity and seismic refraction results respectively.

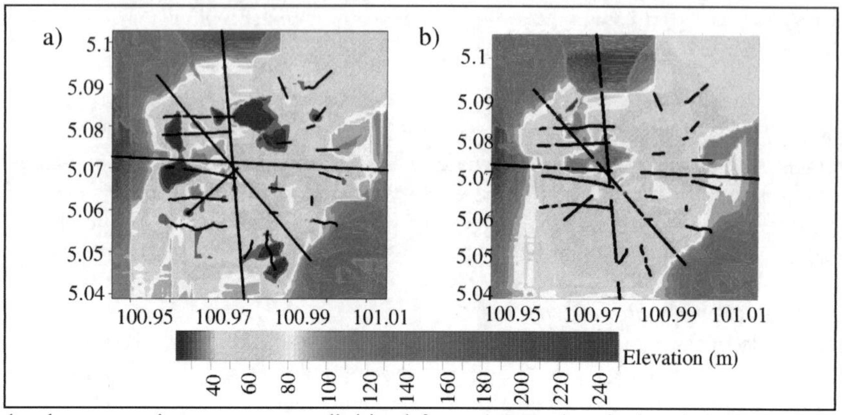

Figure 2. Bedrock topography contour map digitized from a) 2-D electrical resistivity and b) seismic refraction.

Generally, results obtained from 2-D electrical resistivity and seismic refraction methods show a lot of fractures detected within the survey lines inside the crater as well as on the rim. This is suggested due to high impact of meteorite inside the crater and extensive inward collapse around the rim. However, minor fractures detected at survey lines outside the crater. Figure 3 (a, b and c) and figure 4 (a, b and c) shows the 2-D electrical resistivity and seismic refraction sections for area inside, on the rim/slumped terrace and outside the crater respectively.

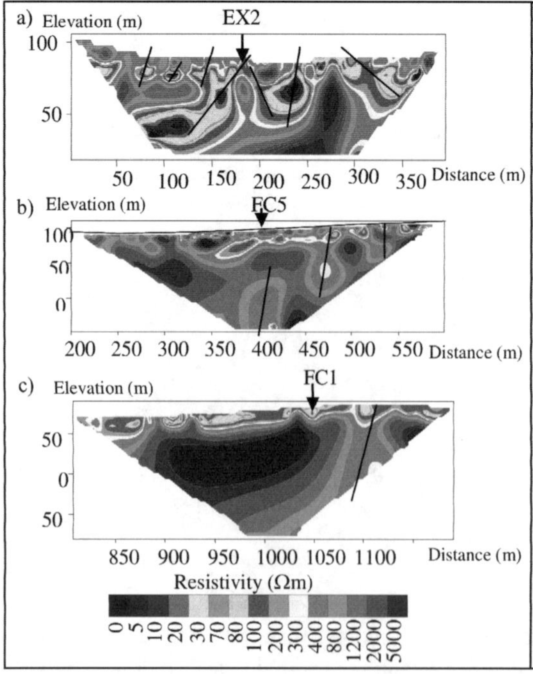

Figure 3. 2-D electrical resistivity section for a) inside crater, b) on the rim/slumped terrace and c) outside crater

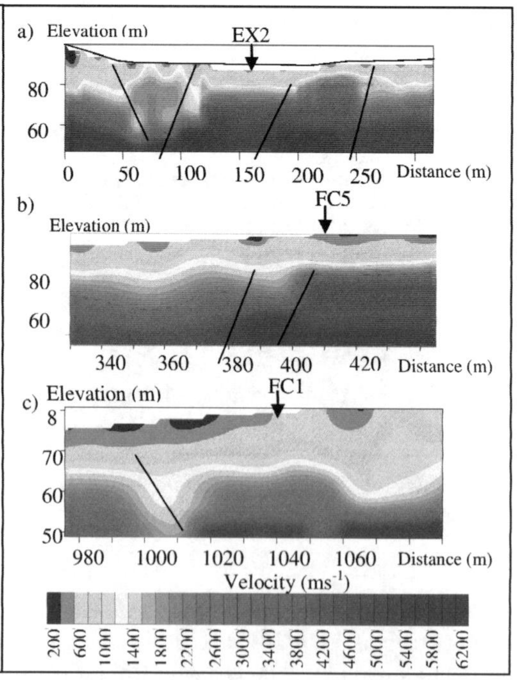

Figure 4. Seismic refraction section for a) inside crater, b) on the rim/slumped terrace and c) outside crater

Results from 2-D electrical resistivity and seismic refraction were correlated with boreholes data that inline and nearest to the survey lines. Standard table of bulk resistivity and p-wave seismic velocity against SPT N value, moisture content and RQD is produce for impacted soils and rock inside the crater, on the rim/slumped terrace and outside crater are generated and presented in Table 1-3 respectively.

Table 1. Impacted soil and rock standard table for inside crater of Bukit Bunuh impact crater

Geological classification	Resistivity, ρ (Ωm)	P-wave velocity, V_p (ms^{-1})	N value	Moisture content, MC	RQD (%)
Post-impact sediment fill deposit					
- clay and silt	100-700	375-800	0-24	18-59	
- sand and gravel	300-5000	800-2100	10-23	12-27	
Rocks					
- Slightly weathered granite					
Class C	1050-2500	1500-2500			70-100
Class D	900-5800	1200-2700			27-50

Table 2. Impacted soil and rock standard table on rim/slumped terrace of Bukit Bunuh impact crater

Geological classification	Resistivity, ρ (Ωm)	P-wave velocity, V_p (ms^{-1})	N value	Moisture content, MC	RQD (%)
Post-impact sediment fill deposit					
- silt	70-500	400-800	2-39	17-30	
- sand and gravel	540-3150	900-3600	10-50	14-26	
Rocks					
- Highly weathered granite	290-530	3200			0
- Moderately weathered granite	250-620	1800-3300			0
- Slightly weathered granite	330-500	1700-3100			17-86
Class D					

Table 3. Impacted soil and rock standard table for outside crater of Bukit Bunuh impact crater

Geological classification	Resistivity, ρ (Ωm)	P-wave velocity, V_p (ms^{-1})	N value	Moisture content, MC	RQD (%)
Post-impact sediment fill deposit					
- silt	55-60	650-700	16-19	18-22	
- sand and gravel	100-420	740-1100	17-50	7-19	
Rocks					
- Slightly weathered granite	1545-1600	2100-2200			0
Class C	870-1150	1500-1900			67-77
Class D	650-700	1260-1300			91.6

Generated standard table for Bukit Bunuh ancient impact crater clearly shows that resistivity value of cohesive soils in overburden layer varies from 100-700 Ωm for both inside crater and on the rim/slumped terrace. While for non-cohesive soils, resistivity value is ranging from 300-5000 Ωm. P-wave velocity of 375-800 ms^{-1} and 800-3600 ms^{-1} is detected for cohesive and non-cohesive soils respectively. Non-homogeneous overburden is identified which mainly due to the high impact of meteorite inside the crater, ejection of basement rocks and fallback of the ejecta. The occurrence of impact melt sheets and suevite breccia also influence the non consistence value of resistivity and p-wave velocity of the overburden. For outside crater area, overburden consists of silt and sand with little gravel has uniform resistivity and velocity distribution. Silt has low resistivity and velocity of 55-60 Ωm and 650-700 ms^{-1} while sand has slightly higher resistivity and velocity value of 100-420 Ωm and 740-1100 ms^{-1} respectively.

Slightly and highly weathered granite bedrock is detected with resistivity and velocity of 900-5800 Ωm and 1200-2700 ms^{-1} respectively for inside crater while low resistivity value of 250-620 Ωm is detected for rim/slumped terrace area. Velocity is distributed in range of 1700-3200 ms^{-1} with totally fractured samples of 0 RQD is observed. While for outside crater area, slightly high and uniform resistivity and velocity of 650-1600 Ωm and 1500-2200 ms^{-1} is obtained respectively which represent by slightly weathered granite indicate that the area is not much affected by the meteorite impact. High SPT N value is identified at crater rim and outside of crater with low percentage of moisture content while low SPT N value is observed inside crater with high percentage of moisture content.

6. Conclusion

Quantitative use of geophysical parameter is less straightforward and typically required the use of correlations with geotechnical parameters. As the in-situ behavioral of soils is complex due to totally site-dependent factors, combined analysis of geophysical and geotechnical parameters are generated for Bukit Bunuh ancient impact crater area based on the crater classification respectively. Impact crater is clearly defined as a complex crater type due to circular pattern observed with some rebound at the center.

7. Acknowledgment

Special thanks are extended to USM Geophysics group for their contribution during field data acquisition. Authors also would like to thank to the Centre for Global Archeological Research Malaysia, Universiti Sains Malaysia for sponsoring the project.

8. References

[1] Shoemaker E M 1997 Why study impact crater? *Impact and explosion cratering:Planetary and terrestrial implications* pp 1-110

[2] Dressler B O, Grieve R A F and Sharpton V L 1994 Large meteorite impacts and planetary evolution Geol. Soc. Amer. Spec. paper 293 pp 348

[3] Grieve R A F, Reny G, Gurov E P and Ryabenko V A 1987 The melt rocks of the Boltysh impact crater, Ukraine, USSR *Contrib. Mineral. Petrol* **96** pp 56-62

[4] Mokhtar S 1993 Kajian perbandingan tapak paleolitik Kampung Temelong dengan Kota Tampan dan sumbangannya terhadap kebudayaan zaman Pleistosein Akhir di Asia Tenggara *Malaysia Museum Journal* **32**.

[5] Mokhtar S 2006 Bukit Bunuh, Lenggong, Perak: Sumbangannya kepada arkeologi dan geologi Negara. *Jurnal Arkeologi Malaysia* **19** pp 1-14.

AeroEarth 2014 IOP Publishing
IOP Conf. Series: Earth and Environmental Science **23** (2015) 012014 doi:10.1088/1755-1315/23/1/012014

An analysis of the role of information and communication technology sectors on Japanese national economy from 1995 through 2005: An application of multiplier analysis

Ubaidillah Zuhdi

Department of Industrial Administration, Tokyo University of Science, Noda 278-8510, Japan
School of Business and Management, Institut Teknologi Bandung, Bandung 40132, Indonesia

ubaidillah.zuhdi@gmail.com

Abstract. The purpose of this study is to continue the previous studies which focused on Japanese Information and Communication Technology (ICT) sectors. More specifically, this study aims to analyze the role of ICT sectors on Japanese national economy using simple household income multiplier, one of the analysis instruments in Input-Output (IO) analysis. The analysis period of this study is from 1995-2005. The results show that the sectors did not have an important role on the national economy of Japan on the period. Besides, the results also show that, from the point of view of the multiplier, Japanese national economy tended to stable during the period.

1. Introduction

The importance of Information and Communication Technology (ICT) on the daily lives of people can be easily perceived in recent years. The evidence is almost all people utilize a mobile phone as one of the communication tools. Besides, the use of computer, both in office and home, is another evidence of the condition. ICT also has an important role on the macro aspect. For example, the international trade between two or more countries will go smoother if ICT is applied.

Many previous studies discussed the ICT topic. For example, [1] compared the calculation results generated by simple output multiplier method and Structural Decomposition Analysis (SDA) which the focus was Japanese ICT sectors. Both tools are analysis tool in Input-Output (IO) analysis. His study aimed to get the other perspective regarding the role of sectors on the national economy of Japan from 1995-2005. Using the similar methodology, [2] analyzed the ICT sectors of Indonesia.

[3] analyzed the effects of final demand modifications on the total output of Japanese ICT sectors in the future. He employed demand-pull IO quantity model, one of the analysis instruments in IO analysis, as a tool of calculation. In his study, the modifications only appeared on the sectors. Using the similar methodology, [4] focused on the Indonesian ICT sectors. In their study, however, the modifications appeared on all Indonesian industrial sectors. The study focused on the ICT sectors of Indonesia and used both modifications was conducted by [5]. On the other hand, the study focused on Japanese energy sectors and also used both modifications was done by [6].

The study analyzes the role of ICT sectors on the national economy of specific country using the other multiplier, viewed from above previous studies, however, is still thin. This analysis is important because it can give the deeper understanding regarding the role. This study is conducted in order to fulfill the gap.

The purpose of this study is to continue the previous studies which focused on Japanese ICT sectors. More specifically, this study aims to analyze the role of ICT sectors on Japanese national economy using income multiplier, one of the multipliers in IO analysis. The analysis period of this study is from 1995-2005.

2. Methodology

The methodology of this study is explained as follows. The first step is to describe the data of study. This study uses Japanese IO tables for 1995, 2000, and 2005 as data. These tables consist of 93, 104, and 108 industrial sectors, respectively. The second step is to conduct the process of aggregation for these industries. This process aims to get the compatibility among different periods. The industries are aggregated into 89 sectors through the process.

The third step is to describe the ICT sectors used in this study. These sectors come from the aggregated industries. Table 1 shows these sectors. The fourth step is to conduct the calculation using income multiplier model. [7] described that this multiplier has two types, namely simple and total household income multipliers. They explained that the former model is described as:

$$m(h)_j = \sum_{i=1}^{n} a_{n+1,i} l_{ij} \tag{1}$$

while the latter one is:

$$\overline{m}(h)_j = \sum_{i=1}^{n+1} a_{n+1,i} \bar{l}_{ij} \tag{2}$$

More specifically, $m(h)_j$, $\overline{m}(h)_j$, n, l_{ij}, and $\bar{l}_{ij}$ are the simple household income multiplier for sector j, the total household income multiplier for sector j, sector numbers, a sector-to-sector multipliers matrix, and a sector-to-sector multipliers matrix with respect to households endogenous, respectively. This study uses the former one as a calculation tool. In this study, the households income part on the row of IO tables is represented by compensation of employees. This part is chosen because it consists of (1) wages and salaries, (2) contribution of employers to social insurance, and (3) other payments and allowances [8]. On the other hand, the part on the column is described by consumption expenditure of outside households.

The next step is to conduct the analysis based on the calculation results. The final step is to describe the conclusions of this study and give the suggestions for the further research. One of the objectives of these suggestions is to get the deeper understanding regarding the role of ICT sectors on Japanese national economy.

Table 1. Japanese ICT sectors used in this study.

No.	Sector number	Sector name
1	72	Communication
2	73	Broadcasting and information services
3	80	Advertising, survey, and information services

(Source: [9])

3. Results and analysis

Tables 2, 3, and 4 show the top five Japanese industrial sectors viewed from the value of simple household income multiplier on 1995, 2000, and 2005, respectively. ICT sectors not appear in these Tables. Therefore, from this phenomenon, one can argue that these sectors did not have an important role on the national economy of Japan from 1995-2005.

On the other hand, Figures 1, 2, and 3 describe the values of simple household income multiplier of all Japanese industrial sectors on 1995, 2000, 2005, respectively. The patterns of the values on the periods are explained in Figures 4, 5, and 6. From the information in these Figures, one can argue that the similar pattern appeared from 1995-2005. In other words, during this period, from the point of view of the multiplier, Japanese national economy tended to stable.

4. Conclusions and further researches

This study, as a continuation of previous studies, analyzed the role of ICT sectors on Japanese national economy from 1995-2005. This study employed simple household income multiplier as an instrument of analysis. The results showed that the sectors did not have an important role on the national economy of Japan on the period. Besides, the results also showed that, from the point of view of the multiplier, Japanese national economy tended to stable during the period.

Table 2. Top five Japanese industrial sectors viewed from the value of simple household income multiplier, 1995.

No.	Sector number	Sector name	Simple household income multiplier
1	75	Education	0.83
2	74	Public administration	0.80
3	78	Social security	0.78
4	79	Other public services	0.72
5	65	Road transport (except transport by private cars)	0.72

Table 3. Top five Japanese industrial sectors viewed from the value of simple household income multiplier, 2000.

No.	Sector number	Sector name	Simple household income multiplier
1	78	Social security	0.80
2	75	Education	0.79
3	76	Research	0.72
4	79	Other public services	0.71
5	65	Road transport (except transport by private cars)	0.71

Table 4. Top five Japanese industrial sectors viewed from the value of simple household income multiplier, 2005.

No.	Sector number	Sector name	Simple household income multiplier
1	78	Social security	0.78
2	75	Education	0.78
3	79	Other public services	0.72
4	65	Road transport (except transport by private cars)	0.69
5	76	Research	0.66

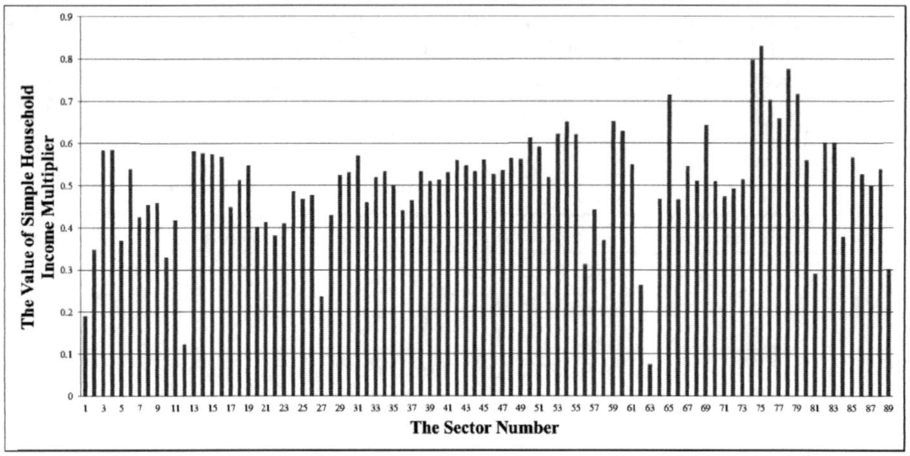

Figure 1. Values of simple household income multiplier of all Japanese industrial sectors, 1995.

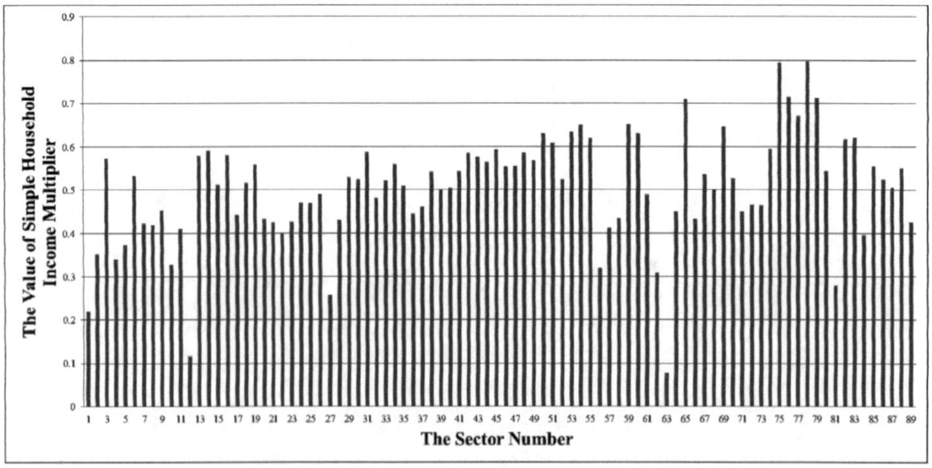

Figure 2. Values of simple household income multiplier of all Japanese industrial sectors, 2000.

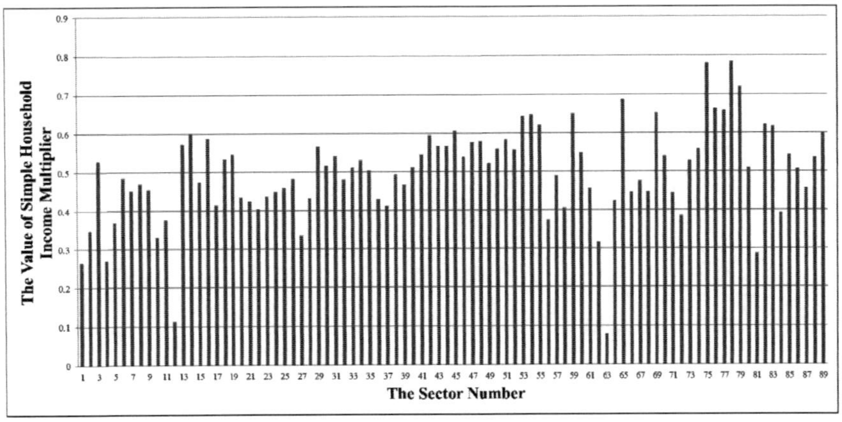

Figure 3. Values of simple household income multiplier of all Japanese industrial sectors, 2005.

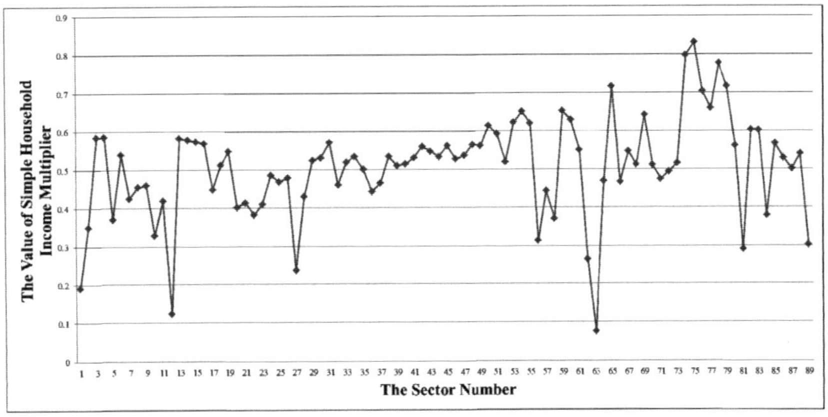

Figure 4. The pattern of values of simple household income multiplier of all Japanese industrial sectors, 1995.

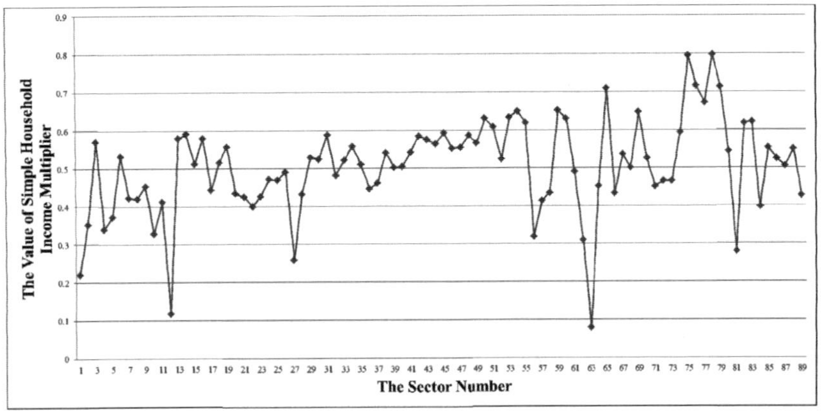

Figure 5. The pattern of values of simple household income multiplier of all Japanese industrial sectors, 2000.

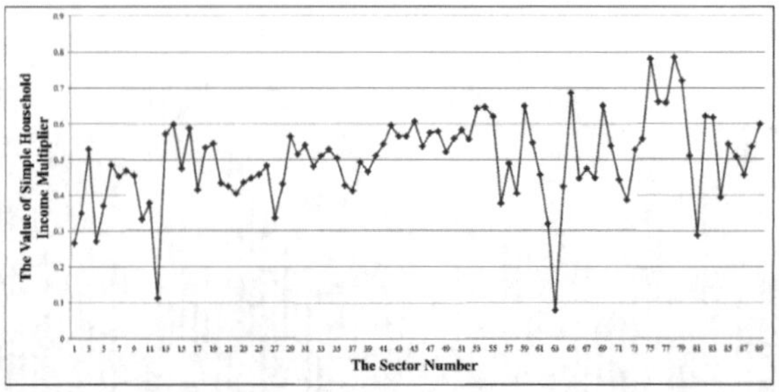

Figure 6. The pattern of values of simple household income multiplier of all Japanese industrial sectors, 2005.

The role of Japanese ICT sectors on the national economy and the patterns of Japanese national economy during the analysis period were discussed in this study. This study, however, did not describe the changes of simple household income multiplier value of these sectors on the period. The discussion regarding these changes is needed in elaborating the characteristics of the sectors during the period. This study suggests this discussion as a further research.

The other suggested further research is to use the value-added multiplier to analyze the role of ICT sectors on the national economy of Japan from 1995-2005. This analysis will explore in more details the contribution of these sectors on Japanese national economy during the period. Besides, this study also suggests the international comparison on the specific analysis period which the compared point is the values of income multipliers of ICT sectors. This comparison will describe in more details the structural changes of these sectors for each analyzed country.

References

[1] Zuhdi U 2014 The other perspective related to the role of ICT sectors in national economy: The case of Japan *Advanced Science Letters* **20** 483486

[2] Zuhdi U 2014 Using multipliers analysis in order to get another perspective related to the role of ICT sectors in national economy of Indonesia: 1990-2005 *Journal of Physics: Conference Series* **495** 012051

[3] Zuhdi U 2014 Analyzing the impacts of final demand changes on total output using input-output approach: The case of Japanese ICT sectors *IOP Conference Series: Earth and Environmental Science* **19** 012016

[4] Zuhdi U, Putranto N A R and Prasetyo A D 2014 Encouraging information and communication technology sectors using input-output approach: The case of Indonesia *Advanced Science Letters* **20** 199202

[5] Zuhdi U 2014 The impacts of final demand changes on total output of Indonesian ICT sectors: An analysis using input-output approach *IOP Conference Series: Materials Science and Engineering* **58** 012011

[6] Zuhdi U 2014 An input-output approach to analyze the ways to increase total output of energy sectors: The case of Japan *IOP Conference Series: Earth and Environmental Science* **19** 012015

[7] Miller R E and Blair P D 2009 *Input-Output Analysis: Foundations and Extensions* (Cambridge: University Press)

[8] Japanese Ministry of Internal Affairs and Communications nd *2005 Input-Output Tables for Japan* available at http://www.stat.go.jp/english/data/io/io05.htm (accessed on September 2014)

[9] Zuhdi U, Mori S and Kamegai K 2012 Analyzing the role of ICT sector to the national economic structural changes by decomposition analysis: The case of Indonesia and Japan *Procedia-Social and Behavioral Sciences* **65** 749754

IOP Conf. Series: Earth and Environmental Science **23** (2015) 012015 doi:10.1088/1755-1315/23/1/012015

An application of multiplier analysis in analyzing the role of information and communication technology sectors on Indonesian national economy: 1990-2005

Ubaidillah Zuhdi

Department of Industrial Administration, Tokyo University of Science, Noda 278-8510, Japan
School of Business and Management, Institut Teknologi Bandung, Bandung 40132, Indonesia

ubaidillah.zuhdi@gmail.com

Abstract. The purpose of this study is to continue the previous studies which focused on Indonesian Information and Communication Technology (ICT) sectors. More specifically, this study aims to analyze the role of ICT sectors on Indonesian national economy using simple household income multiplier, one of the analysis tools in Input-Output (IO) analysis. The analysis period of this study is from 1990-2005. The results show that the sectors did not have an important role on the national economy of Indonesia on the period. Besides, the results also show that, from the point of view of the multiplier, Indonesian national economy tended to stable during the period.

1. Introduction

In recent years, the need for the Information and Communication Technology (ICT) on the daily lives of people can be easily observed. The evidence is almost all people have and utilize a computer in their lives. They use the technology not only for finishing their tasks but also for entertainment purposes. Besides, the use of mobile phone as one of the communication tools is another evidence of the condition. ICT also has an important role on the macro aspect. For example, the activities of political survey will be easier if this technology is applied.

Many previous studies discussed the ICT topic. For example, [1] did the comparison between the calculation results generated by simple output multiplier method and Structural Decomposition Analysis (SDA) which the focus was ICT sectors of Japan. Both tools are analysis instrument in Input-Output (IO) analysis. From the study, he tried to get the other perspective regarding the role of sectors on the national economy of Japan from 1995-2005. Applying the similar methodology, [2] analyzed Indonesian ICT sectors.

[3] analyzed the impacts of final demand changes on the total output of Japanese ICT sectors in the future. He employed demand-pull IO quantity model, one of the analysis tools in IO analysis, as an instrument of calculation. In his study, the modifications only appeared on the sectors. Using the similar methodology, [4] focused on the ICT sectors of Indonesia. In their study, the changes appeared on all Indonesian industrial sectors. The study focused on Indonesian ICT sectors and applied both

modifications was conducted by [5]. On the other hand, the study focused on Japanese energy sectors and also used both changes was done by [6].

The study investigates the role of ICT sectors on the national economy of a particular country using the other multiplier, viewed from above previous studies, however, is still thin. This investigation is important because it can generate the deeper understanding regarding the role. This study is conducted in order to fulfill the gap.

The purpose of this study is to continue the previous studies which focused on the ICT sectors of Indonesia. More specifically, this study aims to analyze the role of ICT sectors on Indonesian national economy using income multiplier, one of the multipliers in IO analysis. The period of analysis of this study is from 1990-2005.

2. Methodology

The methodology of this study is described as follows. The first step is to explain the data of study. This study uses Indonesian IO tables for 1990, 1995, and 2005 as data. These tables consist of 161, 172, and 175 industrial sectors, respectively. The second step is to conduct the process of aggregation for these industries. This process aims to get the compatibility among different periods. The industries are aggregated into 159 sectors through the process. The process of adjustment to get the proper data is addressed to the 2005 IO table of Indonesia before conducting the aggregation procedure.

The third step is to describe the sectors of ICT used in this study. These sectors come from the aggregated industries. Table 1 shows these sectors. The fourth step is to conduct the calculation using the model of income multiplier. [7] described that this multiplier has two types, namely simple and total household income multipliers. They explained that the former model is described as:

$$m(h)_j = \sum_{i=1}^{n} a_{n+1,i} l_{ij} \qquad (1)$$

while the latter one is:

$$\overline{m}(h)_j = \sum_{i=1}^{n+1} a_{n+1,i} \bar{l}_{ij} \qquad (2)$$

More specifically, $m(h)_j$, $\overline{m}(h)_j$, n, l_{ij}, and $\bar{l}_{ij}$ are the simple household income multiplier for sector j, the total household income multiplier for sector j, sector numbers, a sector-to-sector multipliers matrix, and a sector-to-sector multipliers matrix with respect to household endogenous, respectively. This study uses the former one as a calculation tool. In this study, the household income part on the row of IO tables is represented by wage and salary. On the other hand, this part on the column is described by consumption expenditure of household and non-profit institutions.

The next step is to conduct the analysis based on the results of calculation. The final step is to describe the conclusions of this study and give the suggestions for the further research. One of the reasons of proposing the suggestions is to open the opportunity to get the deeper understanding regarding the role of ICT sectors on the Indonesian national economy.

Table 1. Indonesian ICT sectors used in this study.

No.	Sector number	Sector name
1	135	Construction and installation on electricity, gas, water supply, and communication
2	146	Communication services

(Source: [8])

3. Results and analysis

Tables 2, 3, and 4 show the top five Indonesian industrial sectors viewed from the value of simple household income multiplier on 1990, 1995, and 2005, respectively. ICT sectors not appear in these Tables. Therefore, from this phenomenon, one can argue that these sectors did not have an important role on the national economy of Indonesia from 1990-2005.

On the other hand, Figures 1, 2, and 3 describe the values of simple household income multiplier of all Indonesian industrial sectors on 1990, 1995, 2005, respectively. The patterns of the values on the periods are explained in Figures 4, 5, and 6. From the information in these Figures, one can argue that the similar pattern appeared from 1990-2005. In other words, during this period, from the point of view of the multiplier, Indonesian national economy tended to stable.

4. Conclusions and further researches

This study, as a continuation of previous studies, analyzed the role of ICT sectors on Indonesian national economy from 1990-2005. This study employed simple household income multiplier as an instrument of analysis. The results showed that the sectors did not have an important role on the national economy of Indonesia on the period. Besides, the results also showed that, from the point of view of the multiplier, Indonesian national economy tended to stable during the period.

Table 2. Top five Indonesian industrial sectors viewed from the value of simple household income multiplier, 1990.

No.	Sector number	Sector name	Simple household income multiplier
1	151	General government	0.95
2	154	Other community services	0.63
3	152	Education services	0.58
4	11	Rubber	0.52
5	140	Railway transport	0.52

Table 3. Top five Indonesian industrial sectors viewed from the value of simple household income multiplier, 1995.

No.	Sector number	Sector name	Simple household income multiplier
1	151	General government	0.76
2	152	Education services	0.67
3	154	Other community services	0.66
4	11	Rubber	0.52
5	127	Musicals instruments	0.51

Table 4. Top five Indonesian industrial sectors viewed from the value of simple household income multiplier, 2005.

No.	Sector number	Sector name	Simple household income multiplier
1	151	General government	0.64
2	154	Other community services	0.60
3	152	Education services	0.60
4	11	Rubber	0.48
5	155	Private motion picture and its distribution	0.45

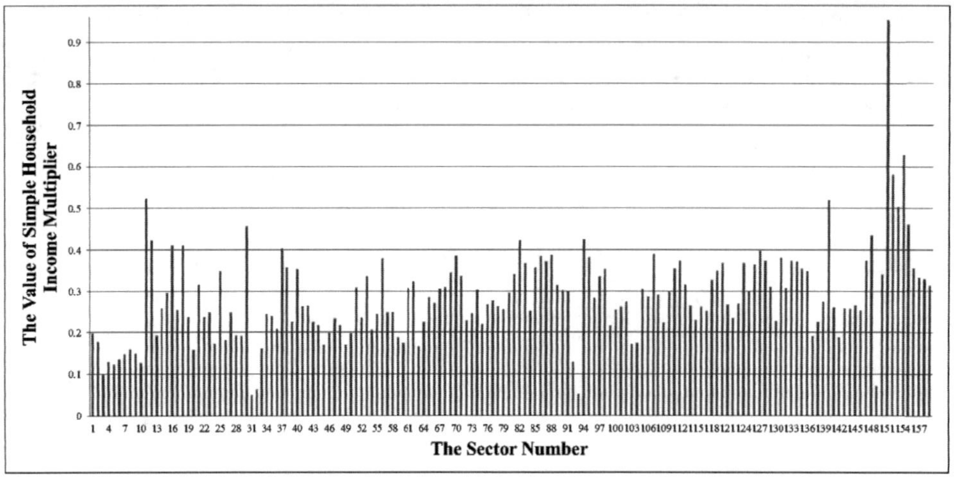

Figure 1. Values of simple household income multiplier of all Indonesian industrial sectors, 1990.

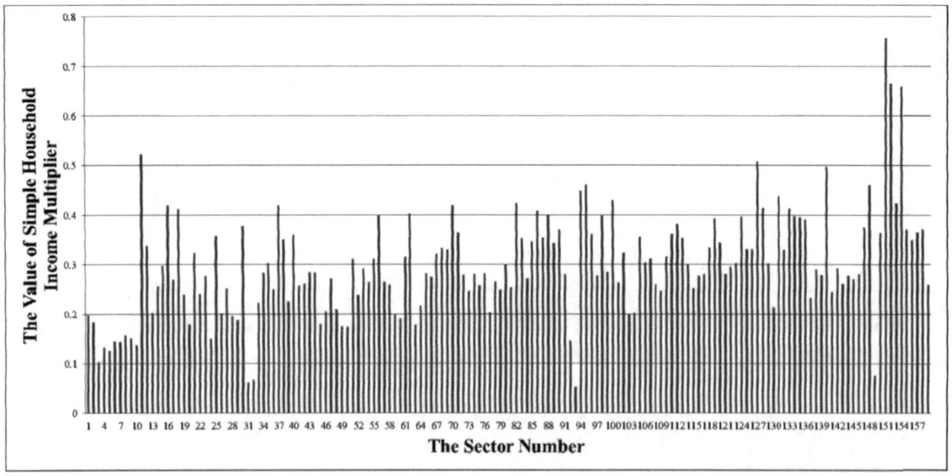

Figure 2. Values of simple household income multiplier of all Indonesian industrial sectors, 1995.

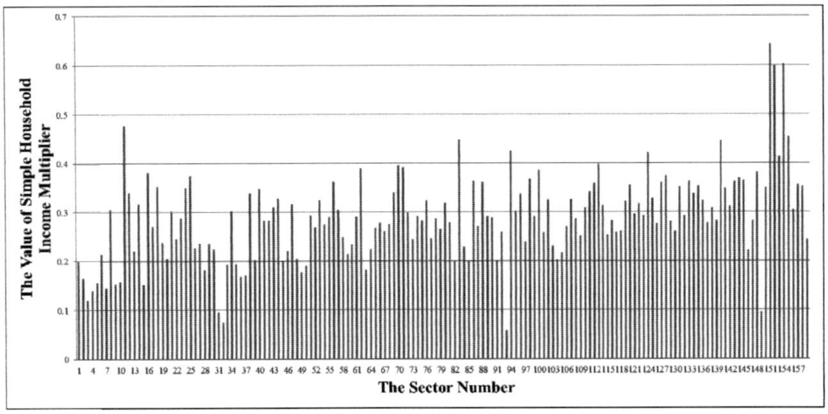

Figure 3. Values of simple household income multiplier of all Indonesian industrial sectors, 2005.

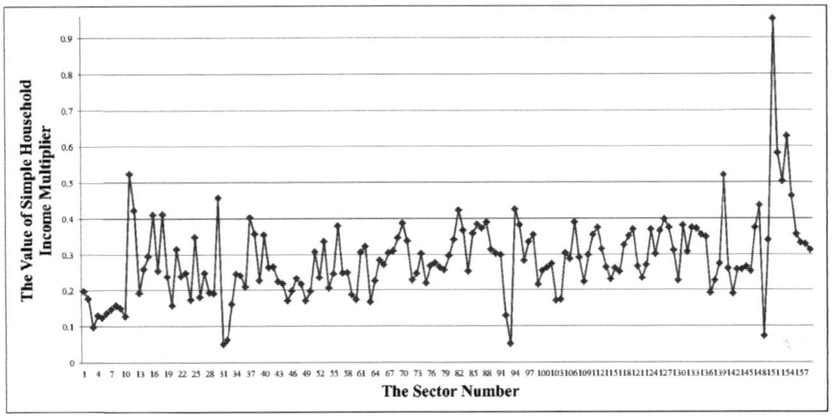

Figure 4. The pattern of values of simple household income multiplier of all Indonesian industrial sectors, 1990.

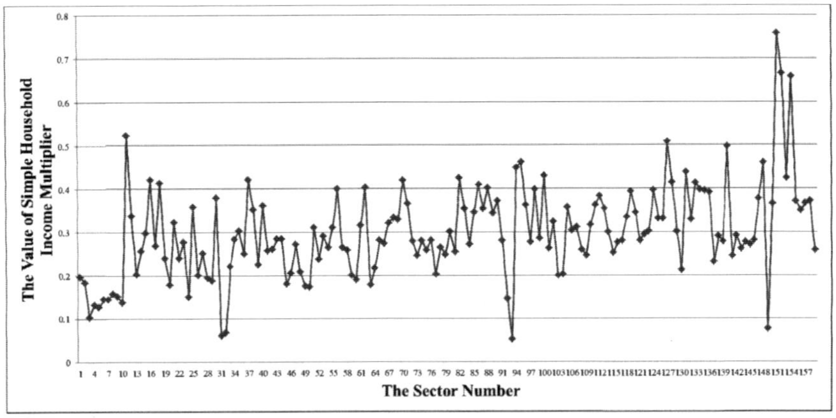

Figure 5. The pattern of values of simple household income multiplier of all Indonesian industrial sectors, 1995.

Figure 6. The pattern of values of simple household income multiplier of all Indonesian industrial sectors, 2005.

The role of Indonesian ICT sectors on the national economy, and the patterns of Indonesian national economy during the analysis period were discussed in this study. This study, however, did not describe the changes of simple household income multiplier value of these sectors on the period. The discussion regarding these changes is needed in elaborating the characteristics of the sectors during the period. This study suggests this discussion as a further research. Besides, expanding the period of analysis of current study is also a proposed further research from this study. This expansion can describe in more details the role of analyzed sectors on and a pattern of the national economy of Indonesia.

The other suggested further research is to use the value-added multiplier to analyze the role of ICT sectors on the national economy of Indonesia from 1990-2005. This analysis can explore in more details the contribution of these sectors on Indonesian national economy during the period. This study also suggests the international comparison on the specific analysis period which the compared point is the values of income multipliers of ICT sectors. This comparison can describe in more details the structural changes of these sectors for each analyzed country. The example of objects for the comparison is developed and developing countries.

References

[1] Zuhdi U 2014 The other perspective related to the role of ICT sectors in national economy: The case of Japan *Advanced Science Letters* **20** 483486

[2] Zuhdi U 2014 Using multipliers analysis in order to get another perspective related to the role of ICT sectors in national economy of Indonesia: 1990-2005 *Journal of Physics: Conference Series* **495** 012051

[3] Zuhdi U 2014 Analyzing the impacts of final demand changes on total output using input-output approach: The case of Japanese ICT sectors *IOP Conference Series: Earth and Environmental Science* **19** 012016

[4] Zuhdi U, Putranto N A R and Prasetyo A D 2014 Encouraging information and communication technology sectors using input-output approach: The case of Indonesia *Advanced Science Letters* **20** 199202

[5] Zuhdi U 2014 The impacts of final demand changes on total output of Indonesian ICT sectors: An analysis using input-output approach *IOP Conference Series: Materials Science and Engineering* **58** 012011

[6] Zuhdi U 2014 An input-output approach to analyze the ways to increase total output of energy sectors: The case of Japan *IOP Conference Series: Earth and Environmental Science* **19** 012015

[7] Miller R E and Blair P D 2009 *Input-Output Analysis: Foundations and Extensions* (Cambridge: University Press)

[8] Zuhdi U, Mori S and Kamegai K 2012 Analyzing the role of ICT sector to the national economic structural changes by decomposition analysis: The case of Indonesia and Japan *Procedia-Social and Behavioral Sciences* **65** 749754

Regional Design Approach in Designing Climatic Responsive Administrative Building in the 21st Century

Hazrina Binti Haja Bava Mohidin[1], Alice Sabrina Ismail

Department of Architecture, Faculty of Built Environment, Universiti Teknologi Malaysia, Skudai. 81310, Johor, Malaysia

E-mail: hazrinahaja@gmail.com

Abstract. The objective of this paper is to explicate on the study of modern administrative building in Malaysia which portrays regional design approach that conforms to the local context and climate by reviewing two case studies; Perdana Putra (1999) and former Prime Minister's Office (1967). This paper is significant because the country's stature and political statement was symbolized by administrative building as a national icon. In other words, it is also viewed as a cultural object that is closely tied to a particular social context and nation historical moment. Administrative building, therefore, may exhibit various meanings. This paper uses structuralism paradigm and semiotic principles as a methodological approach. This paper is of importance for practicing architects and society in the future as it offers new knowledge and understanding in identifying the suitable climatic consideration that may reflect regionalist design approach in modern administrative building. These elements then may be adopted in designing public buildings in the future with regional values that are important for expressing national culture to symbolize the identity of place and society as well as responsive to climate change.

1. Introduction

This paper seeks to elucidate the built form composition of the highest administrative architecture in Malaysia. This study is important because present administrative architecture much portrays eclectic style that incorporates the Greek and Roman, Persian and the Malay vernacular. This mixture of styles caused local identity and values through design lack regional characteristics that suit the local climate and context.

Although there were published writings on administrative building done in the past by architects and academicians like Lai Chee Kien (2007), Izani et al (2009), Rasdi (2001), Ismail (2008), Endut (2001), Vlatseas (2001), Kosman and Nik Ibrahim (2007), Moser (2009), King (2008), Johan Ariffini (2003) and Ho (2006), their works did not include any analysis of the overall administration building from the aspect of façade design or how they should portray regional design approach responsive to local climatic condition. In sum, a survey of the published sources reveals that there is a gap in the literature of the administrative building in Malaysia. To address this gap and to achieve the objective of the study, this paper focuses on documenting regional design characteristics on administrative architecture to produce suitable guidelines in designing better administrative building with consideration to regional characteristics in the future.

2. Background Study

2.1. Defining the Administrative Architecture and arising design issues relating to administrative building in Malaysia

[1] To whom any correspondence should be addressed.

An administrative architecture is a building that accommodates the work of administrators despite their levels to perform all related administrative activities, apart from serving as a place to run the seat of the government. The building is also located in important areas as the capital and houses the office of the government's lawmakers [1]. Administrative buildings are usually built not only to symbolize power and authority of the government, but also to portray the identity of the society at country, states, region, district and communal level to the outside world [1, 2]. At country level, the most important type is the Prime Minister's Office or also known as the capitol complex [1].

However, in the context of Malaysia, many public buildings including the administrative building portray less sensitiveness to design approach that is responsive to the local climatic context. For example, most of the country's administrative building were designed to fit the patrons intend which is usually biased towards the visual impressiveness instead of focusing more on the user comfort. It is heartbreaking to see that Malaysian architecture adapt less to its natural context even if it is abundance in natural resources. Since the highlight of this paper is on the issue of responsive administrative architecture design, it is worthwhile to understand and define the meaning of regional approach design before delving into the two case studies followed by suggestions on how to improve the quality of building design that responds to local climate for future reference.

2.2 Regional approach design and attributes

Regional architecture became a topic that was debated since early modernist era. The term regionalism is defined as a set of associated conditions [3, 4]. Regionalism in the built environment takes into account the practice of architecture within the same doctrine, system, principles such as a municipality or even a small area that shares the same geographical, social, economic and cultural factors. In other words, there are three main components that contribute to the identification of regional values in built form. They are; geography (topography, land, landscape, and climate), local architecture (technology, structural and construction material as well as method of construction) and lastly, local culture (customs, traditions and manners) [4, 5]. Regionalism became favorable and soon the modernist knowledge spread worldwide, which resulted in regionalism to be transplanted globally, even in the tropical regions such as in Malaysia during the late 19th century [6, 7, 8].

2.3 Architectural attributes in determining regional design approach

The reserach on regional climatic approach on administrative architecture started when indoor comfort is being negotiated which results to non-conducive environment to the occupant of the building. During the planning stage, two case studies were selected based on the same architectural typology; government administrative builidng. Research was conducted on the exterior facade of the builidng to study the facade elements. Based on the findings, in order to produce built form that portrays sensitiveness to local climate, many scholars and designers agreed that there are four main architectural elements that should be primarily considered in order to achieve a comfortable indoor occupancy passively. They are – a) roof, b) wall, c) opening, and d) landscape [9 - 11]. However, it was agreed that high-energy consumption as well as indoor comfort are dependent on building façade as it is primarily the main component that sets the tone to the building [11]. Hence for the benefit of this study, only façade elements like roof, wall and opening will be discussed, as it is the most important architectural element that acts as an envelope for the building. To increase the value of the façade design that complies with regional needs, past scholars outlined that exterior building façade should minimizes internal heat gain during daytime while decreases direct solar radiation. This could be achieved by providing many shades from sunlight, rain and water runoffs, as well as emphasis on the usage of regional material for the outer skin [9, 11]. Alongside to this, the building envelope should be affordable, durable, energy-positive, environmentally sensitive, healthy and comfortable as well as intelligent [10]. There are three basic characteristics that promote regional façade design that conforms to regional condition in a hot and humid climate. These basic characteristics are; a) energy efficiency, b) façade types, and c) materials [9-11]. Below are the outlined attributes in determining a building with regional design approach based on studies conducted by [7, 9-11]

Table 2.1. Strategies in responsive design

Characteristics	Key Concept	Strategies
a. Energy Efficiency Ways of minimizing mechanical aids in design	**Reduction of External Heat Gain** protection of the facade from direct solar radiation through self-shading methods (building form) or shading devices	• Large and high roof either domed or pitched • Verandah to cool off building • Large openings to allow air movement
	Cooling use of natural ventilation where environmental characteristics and	• vented to discharge hot air accumulation beneath the ceiling • Double glazed windows to reflect unwanted heat

	building function permit	from entering interior building • Intelligent fenestration – allowing cool air while shading • Full height windows
	Daylight use of natural light sources while minimizing solar heat gain through use of shading devices and light shelves	• Use of skylight to maximize natural light infusion • Vertical and horizontal elements create natural shading or shadowing • Punched or recessed wall • Adequeate placement of openings for maximum air-flow
b. Façade Types Passive approach in achieving indoor comfort	**Solar Control** protection of the facade from direct solar radiation through self-shading methods (building form) or shading devices	• Wide overhangs • Tilted (25°C) façade to admit more desirable light diffusion into building • Tiered or staggered • Shaded – punched or recessed windows
	Thermal and Moisture Resistant use of passive skin approach to protect building from eroding	• Doubel skin façade – cavity wall to create double thermal and moisture barrier
c. Materials Material options without compromising economy, utility, durability, and comfort	**Lightweight material** low thermal capacity materials to keep building cool	• Wood and other natural material that has low thermal capacity • Use of natural material insulator
	Resource Efficiency consideration of building's life-cycle from design, construction, operation, maintenance, renovation, and demolition.	• Natural resources obtained locally • Long life-cycle • Low cost of handling • Easily maintained • Environment friendly • Easily constructed material – extraction, process, transport, install, operation, erection
	Color surface of facade	• Lighter color reflect heat while darker color absorb more solar radiation

3. Method of Study

In trying to explain and describe how the regional design approach is adopted in the two chosen case studies, the administrative buildings are studied in a qualitative manner. This is conducted by giving emphasis on the aspect of building communication since it involves the study of meaning and building as a sign. Due to this reason, structuralism as research paradigm and semiotics as research methodology is adopted since they help to explain how built form may convey meaning through the portrayal of various architectural elements [12, 13]. Due to the limitation of the study, only the external façade of administrative buildings will be used as the basis of this study. This research will adopt direct observation as a method for data collection and layering theory as a basis of analyzing the building skin [12, 14].

3.1 Research methodology

Semiotics, in its most basic definition is the study of signs [13]. The theory of semiotic was developed by scholars such as Roland Barthes and Mark Gottdiener. It is claimed that semiotics exists in a triadic notion. The systems of signification (relationship between the signifier and signified) are multileveled structures, which not only contain denotative signs but also connotative signs when particular cultural codes are ascribed to these signs. Architecture can be fitted into the triadic notion of understanding semiotics as it includes the relationship of the signifier and the signified in a multilevel structure [13]. Therefore, for the benefit of this paper, this theory of semiotics is adapted as it will help to describe and define the meaning of the administrative architecture external facade as a meaningful 'sign.' Data collection will involve direct observation of external façade treatment that will concern roof, wall, and openings. The process of analysis will be further clarified in Table 3.1 below.

3.2 Method of data collection

Direct observation enables diverseness and vastness of context. In this case, direct observation is beneficial to the study because administrative buildings chosen for this study is restricted in its political nature since these buildings are the most important administrative buildings in Malaysia.

IOP Conf. Series: Earth and Environmental Science **23** (2015) 012016 doi:10.1088/1755-1315/23/1/012016

Table 3.1 Method of data collection

Building Indicator	Theory To Analyze	Process of Analysis
Roof, Wall and Openings	Layering technique to analyze the composition of façade elements [14]	1. Observe and identify architectural elements on overall facade 2. Use layering system is to identify elements of form with the following steps: Layer 1: the main form Layer 2: additional built form Layer 3: basic primary elements Layer 4: additional and secondary elements Layer 5: constructive and decorative details Layer 6: entire range of stylistic features

4. Findings

4.1 Background Study

4.1.1. Perdana Putra (1999), Putrajaya. Completed in 1999 on a 5 acre land and cost RM270 million with only 10% imported materials incorporated into the design due to the 1997/1998 economic crisis, Perdana Putra was built to complement the federal administrative territory of Putrajaya. All the consultants were locally handpicked, and the principal architect in charged was Ar. Ahmad Rozi Abd Wahab from AQIDEA.

4.1.2 *Former Prime Minister's (PM's) Office (1967), Kuala Lumpur.* In September 1967, the former PM's office building was officially opened for operation. The office building is three storeys in height and sits on a 0.25-acre land. The cost of construction was RM 1 million using sandblasted concrete as its main façade finishes and structured on reinforced concrete post and beam. The Public Works Department did the overall design and construction of the building while Tunku and the architect initiated most of the design ideas. The former PM's office had accommodated four Prime Ministers before Putrajaya took over the administrative role.

4.2 Architectural Façade Analysis (Layering technique)

4.2.1 Perdana Putra (1999), Putrajaya

Table 4.1 Architectural façade layering of Perdana Putra

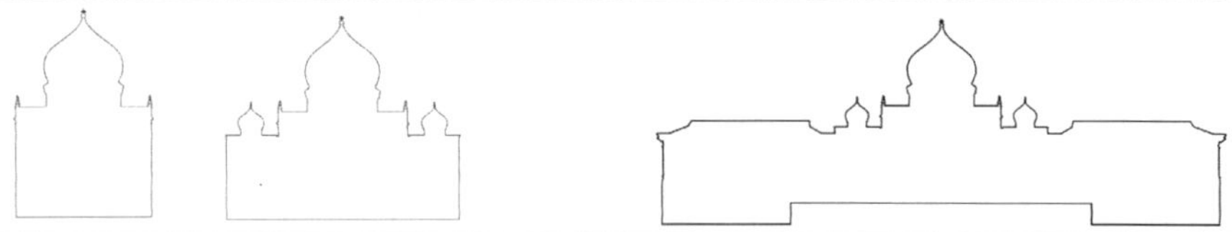

Layer 1 (main form with no hierarchy) Layer 2 (main and additional form)

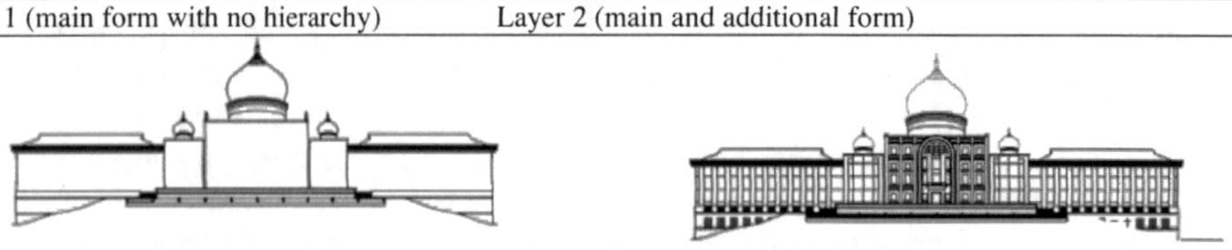

Layer 3 and Layer 4 (Unsymmetrical division of solid and void at façade)

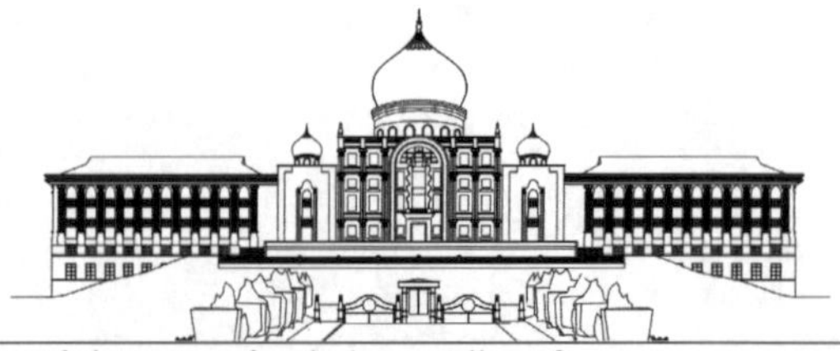

Layer 5 (Simple division of elements on façade: base; wall; roof)

Façade element	
Layer 6 (Simple use of detailing and elements on façade)	

4.2.1.1. Roof. Since roof is the main sheltering element of a building, its primary function is to shade the building as a whole. Perdana Putra uses two types of roof. These roofs are pitched on both ends while domed at the center of the façade. The pitched roof selected for the left and right wing of the building symbolizes the Malay architecture feature where the roof is designed to have a certain length of overhang to shade the immediate wall from solar radiation and rain as well as water runoff. As pitched roof is made of metal roofing that is a non-reflective material and has a greater height in the middle, it is logically more effective in creating less hot air pocket under the ceiling as the height has been increased. The same idea was adapted on the Indo-Saracenic and Moguls influenced green domed roofing and this ensures better indoor air quality. However, the material used for the domed roof; green stone is rather reflective and can easily ignite thermal conductivity. The light green color roofing reduces heat gain as well as reducing glare.

4.2.1.2. Wall. The wall of Perdana Putra is composed of large arches that create recessed wall and helps to shade the double-glazed full sized window. These arches are adaptation of Baroque and Neo-classicism architecture style that originated from France. The wall of Perdana Putra created a sense of interaction between interior and exterior space that was idealized by adopting concrete wall façade with openings that are full in height. It is also embellished with Classical ordered - Ionic columns that create further recess on the façade. The wall uses granite stone as one of the main materials. These granite stones are lightly colored, and its surface is not shiny that helps to reduce the heat gain. The availability of a verandah on the center building helps to reduce the cooling load by minimizing usage of mechanical services.

4.2.1.3. Opening The large opening of Perdana Putra are full in height and are embellished with wrought iron grill detailing that originates from a combination of Islamic arts and art nouveau architecture style. These styles are clearly seen on the large center arch that is heavily emphasised. The recessed double glazed windows reduce solar radiation while allowing natural daylighting as well as air to enter the inside spaces of the building. There are also opening on the dome's sanctuary to help reduce hot air that rises from the bottom to top by means of stack effect.

4.2.2 Former Prime Minister's Office (1967), Kuala Lumpur

Table 4.2 Architectural façade layering of Former Prime Minister's Office

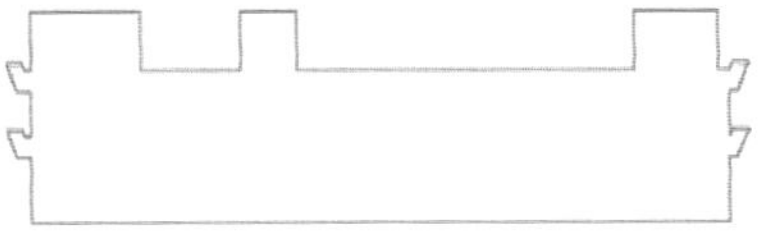	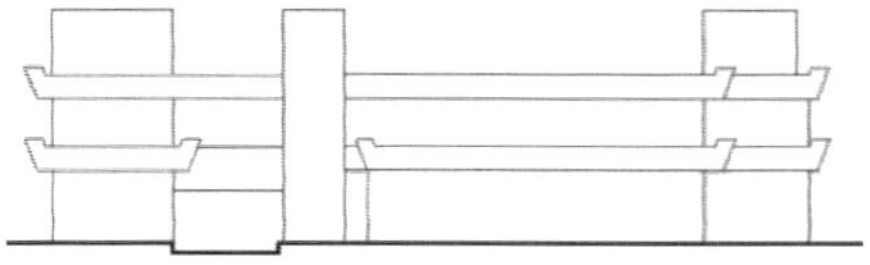
Layer 1 (main form with no hierarchy)	Layer 2 (main and additional form)
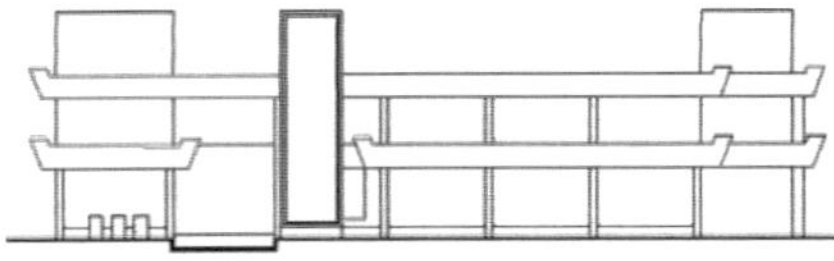	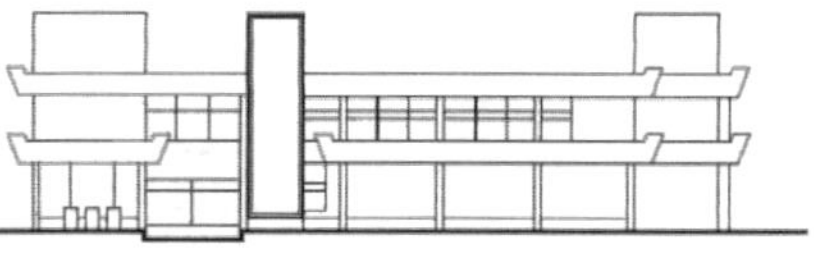
Layer 3 and Layer 4 (Unsymmetrical division of solid and void at façade)	

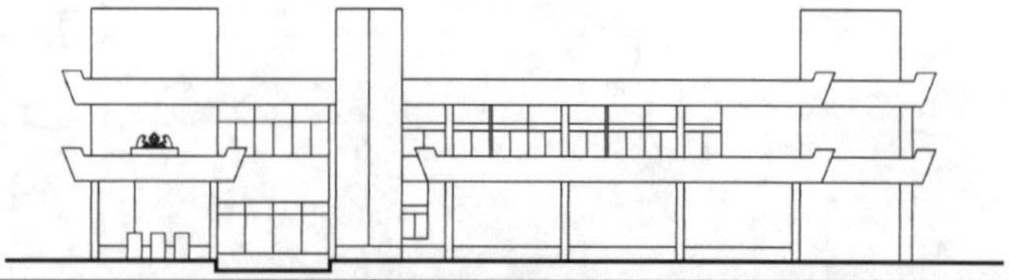

Layer 5 (Simple division of elements on façade: base; wall; roof)

Façade elements

Layer 6 (Simple use of detailing and elements on façade)

4.2.2.1 Roof. The former Prime Minister's (PM's) office has a flat roof covering the building as a whole. It is being designed in a staggered and tiered manner that shades not only the immediate wall and openings but also shelters all floor levels as well. Although being flat in nature, the roof of this building had a long overhang and cantilevered at all façade of the building creating a recessed wall that could shade better and helps to reduce the usage of mechanical cooling system within the interior spaces. The roof material is of non-reflective source that does not cause thermal conductivity. Being brutalist-expressionism in style, the roof of this building is made of bare and exposed materials using abrasive sandblasted concrete and stucco.

4.2.2.2 Wall. The wall of former PM's office is rich in its horizontal and vertical elements that create one or more ways to adapt better in the region's climatic condition. These horizontal elements gave a deep recess to the wall and this result to a better shading solution on every floor level. The wall treatments are made of the same material used consistently throughout the façade. Sandblasted concrete and stucco are the two choices of material that was being considered to be effective in keeping heat during night times while resisting heat load during the day. However, these materials are dark in color making it better in heat absorption.

4.2.2.3 Openings. The interior and exterior spaces of the building are modeled to have a sense of interaction where large windows are placed on its façade. These windows are also recessed to allow good daylighting penetration while eliminating glare and solar radiation. It is also rain reflective and avoids water run-offs at each opening, and at the same time this could reduce the formation of mould as an effect of high humidity level. There is also a high degree of protruding façade on its main opening, giving good protection against the weather as well good air circulation and distribution.

5. Analyses and Discussion

Table 5.1 architectural façade analyses on two administrative architecture in Malaysia

Façade Elements	Required Characteristics (Guidelines)	Perdana Putra	Former PM's Office
Roof	• Large and high roof either domed or pitched	/	
	• vented to discharge hot air accumulation beneath the ceiling	/	
	• Use of skylight to maximize natural light infusion		
	• Wide overhangs	/	/
	• Lightweight material	/	/
	• Resource Efficiency	/	/
	• Light colored surface	/	
Wall	• Verandah to cool off building	/	
	• Double glazed windows to reflect unwanted heat	/	/
	• Vertical and horizontal elements create natural shading or shadowing	/	/
	• Punched or recessed wall	/	/
	• Tilted (25 ℃) façade to admit more desirable light diffusion into building		
	• Tiered or staggered		/

	• Double skin façade – cavity wall to create thermal and moisture barrier		
	• Lightweight material	/	/
	• Resource Efficiency	/	/
	• Light colored surface	/	
Openings	• Large openings to allow air movement	/	/
	• Intelligent fenestration – allowing cool air while shading		
	• Full height windows	/	/
	• Adequate placement of openings for maximum air-flow	/	/
	• Shaded – punched or recessed windows	/	/
	• Lightweight material	/	/
	• Resource Efficiency	/	/
	• Light colored surface		

It can be seen that Perdana Putra adapts more to the climatic condition in all façade elements; roof, wall and openings. As compared to former PM's office, Perdana Putra has a better approach on its roof style. Although, on the overall comment, Perdana Putra seems to be more responsive, it lacks in its wall design approach. The former PM's office has a better wall design approach as the tiered and staggered design enable the building to be shaded naturally. As for the opening, Perdana Putra and the former PM's office are equally responsive based on the façade analysis.

6. Conclusion

Indoor comfort is one of the main factors for a productive employee, and therefore, an administrative building is required to meet responsive design standard especially on façade design as the building envelope. It is highly recommended that future architects and policy makers to follow the guidelines as stated in table 5.1 as an effective adaptation on façade design as façade is the main architectural element that is commonly exposed to harsh climatic conditions and as an intelligent measure in reducing energy consumption loading in administrative building particularly. This paper will also benefit the public at large, being part of the user of the building, as it will educate them to appreciate architectural façade appropriately so as to maximize comfort by optimizing façade elements. The research on regional design approach can be expanded to other architectural elements such as the form of the building and spatial arrangement. The guideline in table 5.1 can also be used as a tool to examine the regional climatic approach on other typological buildings especially public government buildings.

References

[1] Vale, J. Lawrence (2008) *Archtiecture, Power, and National Identity*. London and New York: Routledge.
[2] De Rossi, A. (2009). *Grande SCALA. Architecture Politic Form*. List Laboratorio Internazionale Editoriale.
[3] Dickinson, R. E. (2013). *City, Region And Regionalism: A Geographical Contribution To Human Ecology*. Routledge.
[4] Mursib, G. (2008). *Pemikiran dan sahutan regionalisme dalam wacana senibina Malaysia* (Doctoral dissertation, Universiti Teknologi Malaysia, Faculty of Built Environment).
[5] Marhani, M. A., Adnan, H., & Ismail, F. (2013). OHSAS 18001: A Pilot Study of Towards Sustainable Construction in Malaysia. *Procedia-Social and Behavioral Sciences*,(ajE-Bs) *85*, 51-60. (5)
[6] Bay, J. H., & Ong, B. L. (Eds.). (2007). *Tropical Regional Architecture*. Routledge.
[7] Ahmad, M. H. (2005). *Can Climates Response Become An Essential Lingua Franca For Malay Architecture? International Seminar, Malay Architecture As Lingua Franca*, June 22 & 23, 2005, Trisakti University, Jakarta, Indonesia (
[8] Ismail, A. S. (2002). *Gagasan Pemikiran Frank Lloyd Wright Ke Arah Pembentukan Identiti Seni Bina di Malaysia*. Malaysia: Desktop Publication.
[9] Hyde, R. (2013). *Climate Responsive Design: A Study Of Buildings In Moderate And Hot Humid Climates*. Taylor & Francis.
[10] Haase, M., & Amato, A. (2006, September). Regional Façade Design For Zero Energy Buildings In The Tropics. In Clever Design, *Affordable Comfort–Proceedings Of The 23rd International Conference On Passive And Low Energy Architecture (Plea)*, Geneva, Switzerland.
[11] Ahmad, M. H., Ossen, D. R., & Lim, Y. W. (2008). Review on solar shading strategies and measuring tools for building energy efficiency in tropical climate. *10th SENVAR + 2nd ISESEE 2008 Humanity & Technology*, 2008, Shah Alam.
[12] Ismail, A. S. (2008) *A multidisciplinary approach in understanding the symbolic meaning of built form. In: Methodology in Architectural Research*. Penerbit UTM , Johor, pp. 180-211.
[13] Leeds-Hurwitz, W. (2012). *Semiotics and communication: Signs, codes, cultures*. Routledge.
[14] Shatha, M. (2004). Local identity of modern Amman–Jordan: a perceptual approach towards identifying an inter-subjective and shared architectural schemata. *Urban design international*, *9*(3), 119-131.

Exploring Modular Architecture for Nano Satellite and Opportunity for Developing Countries

M. K. Rhaman [1]*, M. I. Monowar [2], S. R. Shakil [3]*, A. H. Kafi [4], R. S. I. Antara [5]
School of Engineering and Science, BRAC University, 66, Mohakhali, Dhaka-1212, Bangladesh .

E-mail: [1]*khalilur@bracu.ac.bd, [2]maisunmonwar@gmail.com, [3]*shifur.shakil@gmail.com, [4]ahkafi29@gmail.com, [5]antara.anto@gmail.com

Abstract. SPACE Technology has the potential to provide information, infrastructure and inspiration that meets national needs in developing countries like Bangladesh. Many countries recognize this; in response they are investing in new national satellite programs to harness satellite services. Technology related to space is one example of a tool that can contribute to development both by addressing societal challenges and by advancing a nation's technological capability. To cope up with the advanced world in space technology Bangladesh seems to be highly potential country for satellite, Robotics, embedded systems and renewable energy research. BRAC University, Bangladesh is planning to launch a nano satellite with the collaboration of KIT, Japan. The proposed nano satellite project mission is to experiment about social, commercial and agricultural survey needs in Bangladesh. Each of the proposed applications of the project will improve the lives of millions of people of Bangladesh and it will be a pathfinder mission for the people of this country. Another intention of this project is to create a cheap satellite based remote sensing for developing countries as the idea of large space systems is very costly for us therefore we have decided to make a Nano-satellite.

1. Introduction

A number of international organizations, including the United Nations, have recognized the opportunity for space technology to serve developing countries; they pursue long term programs to promote this potential. Meanwhile governments in many developing countries also view that space an important tool for their development. All nations have on-going activities to ensure that space is harnessed in their country. Bangladesh is also establishing programs to operate a nationally owned satellite which is named as BANGABONDHU Satellite. This trend is enabled by the increasing performance of smaller, less expensive satellites. The countries that are the focus of this study seek to transition from owning satellites purchased abroad to attaining national capability to design and build satellites. One approach that countries around the world are using to initiate satellite development capability is to execute a Collaborative Satellite Development Project (CSDP) in partnership with an experienced foreign agency [1]. BRAC University would like to do the same by the collaboration with LaSEINE lab of KIT Japan, JAXA, NASA, ISRO, SPARRSO etc. Collaboration with KIT and SPARRSO is under process. BRAC University has a warm relation with NASA. So, it is very possible to continue this kind of research which will open a new dimension of research in Bangladesh.

The satellite applications are mainly considered for remote sensing, to provide imagery and environmental data; communication and navigation. The primary mission objective is to monitor formation and depression of Bay of Bengal leading to major hurricane, typhoon and other storm. The reason for monitoring monsoon wind & rainfall characteristics is to because excessive rainfall causes flood which is one of the major natural disaster of our country. Meteorologists are reporting that rainfall in Bangladesh in changing drastically (early on set and late on withdrawal in monsoon rain). Moreover, weather forecast use various observations from which to analyses the current state of the atmosphere. As Bangladesh is an agricultural country and most of the people here depend on agriculture for their livelihood therefore agriculture is one of the most important application fields of satellite. By using earth observation data we can do crop inventory, yield prediction, soil/crop condition monitoring and subsidy control. Another advantageous of this nano satellite is to monitor the forestry which helps to balance the eco-system as in recent years killing animals and cutting down trees illegally increasing immensely. By analyzing earth observation data can protect natural resources and ensure the safety of the animals in forest.

2. Implementation Plan

The main target of Bangladesh is to send own satellite in space. It was figured out that the VISION of building small satellites is very possible in five years if we can give emphasis on human resource capacity building & develop an infrastructure. The very first step has been taken by The Robotics lab of BRAC University, Bangladesh to start this project by signing research collaboration with LaSEINE laboratory of Kyushu Institute of Technology, Japan who are pioneer on nano-satellite research in Japan. The project will be led by a team of faculty members from BRAC University supervising a group of undergraduate and postgraduate students.

For project management a full time research assistant will be appointed. This kind of extensive research is very expensive and need high end technology & researcher. Since it also involve the national security and interest, it is very difficult to continue without direct support of Government. So, we are expecting financial help from the government of our country. We plan to deploy our nanoSat from ISS through nanoSat launching Robotic arm. With the collaboration of KIT and JAXA, our plan is to send our nanoSat in a cargo ship such as HTV-IV.

3. Proposed Model

The proposed structure is composed of six individual pyramid shapes whose vertex has been clipped. Each Pyramid is 10 cm x 10 cm at the base with the height of 5 cm. Each pyramid structure has been clipped at 2.5 cm height (Figure 1).

Figure 1. A single pyramid Unit

Assembling six pyramids therefore leave a cavity of 5 cm x 5 cm x 5 cm in the center. This space is mainly intended for battery (Figure 2).

Figure 2. Six pyramids assembled to form a cube.

In contrast to usual cube sat structure, this structure allows us to setup observatory equipments to all six sides. Additionally, as the structure is made of six individual units, we can design and test each mission as a unit. This will further shorten the length of cubesat development period. Our aim is to adopt lego style architecture for cubesat. Of course, this will not allow us to build structure like 2u (10 cm x 10 cm x 20 cm), 3u (10 cm x 10 cm x 30 cm) cubesat format. By increasing the base size of the individual pyramids, the size of the cubesat can be increased.

Each unit has 2 solar panels. The solar panels match the size of the base of the pyramid structure. Two solar panels are deployable and they fold on top of the other as seen in the Figure 3. The units are assembled alternatively, so when the solar panels are deployed they take the form as indicated in this Figure 4.

Figure 3. A single unit with its solar panels folded.

Figure 4. A single unit with its solar panels deployed.

In our design we could stack three circuit boards on top of another (Figure 5). The board sizes are 6.66 cm x 6.66 cm, 8.33 cm x 8.33 cm. Gap between top board and the middle board is 0.835 cm. The gap between lower board and the middle board is 0.88 cm. These boards can either house sensors or control circuits. Top circuit board can be a solar panel, sensor or a camera as per mission needs.

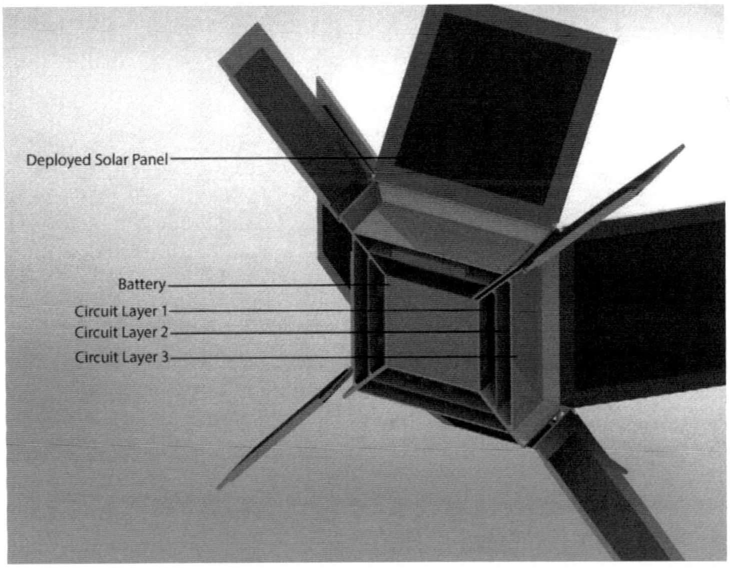

Figure 5. Cross sectional view of assembled nanoSat with solar panels deployed.

Overall, the structure is entirely modular like lego piece (Figure 6). Cubesat can be assembled and later modified as per mission requirement. Trimetric view of proposed model (solar panels undeployed) and trimetric view of proposed model (solar panels deployed) are shown in Figure 7 and 8 respectively.

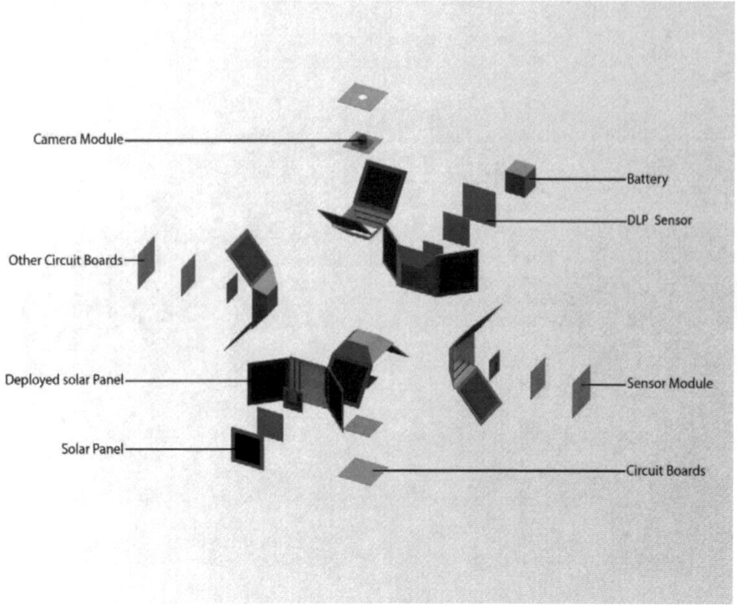

Figure 6. Exploded view of nanoSat.

Figure 7. Trimetric view of proposed model (solar panels undeployed).

Three axis Reaction wheels are very essential for our design as there is a requirement for constant repositioning of nano satellite in order to get the maximum benefit from six open faces. Operating a single axis reaction wheel can take upto 28 watts of power, whereas a typical 1u solar panel can provide two watts of power. Continuous operation of reaction wheel will drain the battery quickly. In addition, it will put increased load on battery, decreasing its longevity. Hence we need to increase the number of solar panels so that it could harness enough power for the equipments to operate properly.

Figure 8. Trimetric view of proposed model (solar panels deployed).

4. Control Mechanism

In low-pressure plasma discharges (<1 Torr), Langmuir probes are commonly used for diagnostics purposes. Under the assumption of local thermodynamic equilibrium and known temperature, Langmuir probes may be used to measure the ion density of an ionized gas, such as a flame at high pressure (>100 Torr) [2]. It is not possible to determine plasma parameters using the Langmuir probe but the plasma current may be determined using a double Langmuir probe along with linear library least squares fitting. The current collected by the probe obviously depends on the three factors: density of the charge carriers in gas, the distribution of their velocity, its shape size [3-5]. In our proposed system we will use this type of Double Langmuir Probe (DLP) sensor to determine the electron density and to diagnosis of plasma. Although the measurements will not be precise, it should provide us with enough information to get students acquainted with space physics.

The main benefits of a GPS subsystem for a small satellite are global cover age and high accuracy at a low cost along with the real-time on board availability of navigation data [6]. The power consumption of GPS receivers vary depending on their type but are usually in the range from 100mW to a few watts [7]. Usually single-frequency receivers are chosen for general navigation applications while dual-frequency receivers are chosen for GPS- related scientific purposes. GPS can also be used for coarse attitude determination [8]. If the direction of the GPS satellite signals can be found, the orientation of the antennas can be found by comparing with the known GPS satellite positions.

Sun sensor is a kind of optical attitude sensor, which is used to get the orientation information of a spacecraft relative to the sun by determining the position of sun vector in the satellite coordinate system through sensing the position of sun vector [9]. When the satellite is in a certain position, the vector of the sun in the satellite orbit coordinate system is S_o, and the output vector of the sun by the sun sensor is S_b, therefore

$$S_b = T_o^b S_o + v$$

Where, v is the Measurement error of a sun sensor and T_o^b

As all of the six faces can have sensors, alignment of the sensors in free space is vital. Thruster will not be able to provide to enough thrust for elongated period of time. Therefore we have to rely on reaction wheel [10]. Reaction wheel allows us to align our cubesat according to our requirement and sensor data through angular momentum. Due to constant requirement for alignment, reaction wheels will consume more power. This difficulty is overcome by the deployable satellites in our model, which should generate enough electricity without needing MPPT. Three reaction wheels should be mounted without distorting load balance of the entire cubesat [11].

5. Communication & Ground System

Due to the unique structure of our nano satellite, it is required to redesign communication module that will send data back to our ground station. The main objective of the satellite is to get the scientific data and to transmit that data back to Earth, so the main challenge is the transmission the data to and from the satellite. For small size satellite it is better to use an array of frequency bands to communicate. Our purpose is to maximize the data transfer rate and minimize the hardware complication. That's why we have to increase carrier signal frequency and increase the data transfer speed. VHF/UHF transmitters are a consistent, low cost solution for data transfer. These systems are naturally used in LEO with Omni-directional antennas, and therefore do not need a high level of pointing accuracy. Transceivers/transmitters in this category can cost from hundreds to a few thousand dollars.

Antenna of onboard transmitter will be directly embedded into the deployable solar panel since solar panels will deploy in all direction. We have the freedom to place antenna on any plane best suited for the given circumstances.

For our ground station, COTS cubeSat products can be readily used. The design of the antenna is being directly related to the transceiver on the nano satellite.

6. Scope of This Research

This type of research work needs a lot of money due to the high material cost and extreme complexity. Moreover, the security of the country is involving with this issue therefore the approval and funding of government is indispensable for this project. The first target group for the proposed research will be the international organizations, in particular United Nations Specialized Agencies based in Geneva (HCR, ILO, ITU, UNOG, WHO, WMO, WTO, etc.)These organizations offer a truly international and multicultural environment.

The second circle of organizations would be the ITU Sector Members, e.g. the private sector company's members of the International Telecommunication Union, and which represent the broader ICT industry. Most of these companies are multinational entities and make a wide usage of ICTs for their own business. Both circles therefore offer a proper environment, corresponding to the framework of the proposed research.

7. Discussion

A project examined trends in the use of satellite-based technologies for needs in developing countries [12]. Their results enable the better understanding of the technology and management options that are being pursued in satellite projects. The content of that outcomes are provided for both developing country policy makers and potential development partners. The satellite applications considered in that research are as remote sensing, to provide imagery and environmental data, communications (including phone, radio, television, and internet) and navigation, which facilitates services such as air traffic management and surveying. Each of these applications can improve the lives of millions of people.

Nano satellite, which is small in volume and weight less than a kilogram is a complex system. The equipment needed for this satellite is not available in Bangladesh and we have to design this satellite by purchasing equipment's from abroad or we have to design satellite compatible sensors, solar panel, power cells, control systems and mechanical structure and make it from expert countries. We have no highly equipped lab for testing all the equipment's of satellite. We have to test all the equipment, sensors, mechanical structure, solar, panel and lastly the whole satellite. As this is the first time we are starting our research to build our own satellite, so obviously we need a lot of help. LaSEINE lab of KIT, Japan is highly equipped with entire testing facilities for the satellite related equipment's including the whole small size nano-satellite. Hopefully with the help of KIT, testing problem will be solved. As nano-satellite is very small in size and in volume so it is really tough to assemble lots of equipment in so tiny place. To find a way of this problem we need also Robotics Experts.

Acknowledgement
The authors are grateful to Prof A A Ziauddin Ahmad, Chairperson, Department of Mathematics & Natural Science, BRAC University (Former Chairman of SPARRSO), Prof. Md. Mosaddequr Rahman, Professor, SECS, BRAC University, Dr. Mohammed Belal Hossain Bhuian, Associate Professor, SECS, BRAC University, Dr. Tarem Ahmed, Assistant Professor, SECS, BRAC University. Dr. John L Polansky and Dr. Arifur khan are acknowledged for their boundless support.

Reference

[1] Danielle RW, Affiliated with The Aerospace Corporation,Chantilly, VA, 20151 (Analysis of Technology Transfer within Satellite Programs in Developing Countries using Systems Architecture)
[2] Cohen IM *The physics of Fluids* 1963 **6** 1492-1500
[3] Kanazawa SMK, Moriwaki T, Okazaki S *J.Phys.D* 1988 **21** 836
[4] Roth JR Industrial Plasma Engineering 1995 **1** 121
[5] Foust, J.: NASA seeks to buy a dedicated nanosatellite launch. [Online] 2014.
Available : http://www.newspacejournal.com/2013/08/14/nasa-seeks-to-buy-adedicatednanosatellite-launch/
[6] Klofas, Bryan: CubeSat Developers' Workshop 2013 [online] 2013. Available at:
http://mstl.atl.calpoly.edu/~bklofas/Presentations/DevelopersWorkshop2013/
[7] Kyohei A., et. al.: High-Speed Attitude Control System for Small Spacecraft with Micro-CMGs. 27th International Symposium on Space Technology and Science (ISTS), Tsukuba, July 5-10, 2009-d-23, 2009.
[8] Lappas V et. al. Micro CMGs for Agile Small Spacecraft: Design and In-Orbit Tests. Proceedings of the 6th International ESA Conference on Guidance, Navigation and Control Systems. Loutraki, Greece, Oct. 17-20, 2005.
[9] Larson WJ, Wertz JR, Space Mission Analysis and Design Microcosm Press, 3rd ed. El Segundo, CA, 2004.
[10] Smith JE, Attitude Model of a Reaction Wheel/Fixed Thruster Based Satellite Using Telemetry Data. No. AFIT/GA/ENY/05-M010. Air Force Inst of Tech Wright-Patterson AFB OH School of Engineering and Management/Dept of Engineering physics, 2005.
[11] Kalman A, Enhanced Power Systems for CubeSats *San Luis Obispo*, 2012.
[12] Danielle RW, Annalisa W, IAC-08-b4.1.1, Massachusetts institute of technology cambridge, massachusetts (The Use of Satellite-Based Technology To Meet Needs In Developing Countries).

AUTHOR INDEX